TM

References for the Rest of Us!®

BESTSELLING BOOK SERIES

Do you find that traditional reference books are overloaded with technical details and advice you'll never use? Do you postpone important life decisions because you just don't want to deal with them? Then our *For Dummies®* business and general reference book series is for you.

For Dummies business and general reference books are written for those frustrated and hard-working souls who know they aren't dumb, but find that the myriad of personal and business issues and the accompanying horror stories make them feel helpless. *For Dummies* books use a lighthearted approach, a down-to-earth style, and even cartoons and humorous icons to dispel fears and build confidence. Lighthearted but not lightweight, these books are perfect survival guides to solve your everyday personal and business problems.

> *"More than a publishing phenomenon, 'Dummies' is a sign of the times."*
>
> — The New York Times

> *"A world of detailed and authoritative information is packed into them..."*
>
> — U.S. News and World Report

> *"...you won't go wrong buying them."*
>
> — Walter Mossberg, Wall Street Journal, on For Dummies books

Already, millions of satisfied readers agree. They have made For Dummies the #1 introductory level computer book series and a best-selling business book series. They have written asking for more. So, if you're looking for the best and easiest way to learn about business and other general reference topics, look to For Dummies to give you a helping hand.

Wiley Publishing, Inc.

5/09

Beekeeping
FOR
DUMMIES®

Beekeeping FOR DUMMIES®

by Howland Blackiston

Foreword by Kim Flottum
Editor, Bee Culture Magazine
President, Eastern Apicultural Society

WILEY

Wiley Publishing, Inc.

Beekeeping For Dummies®

Published by
Wiley Publishing, Inc.
111 River Street
Hoboken, NJ 07030
www.wiley.com

For general information on our other products and services or to obtain technical support, please contact our Customer Care Department within the U.S. at 800-762-2974, outside the U.S. at 317-572-3993, or fax 317-572-4002.

Wiley also publishes its books in a variety of electronic formats. Some content that appears in print may not be available in electronic books.

Library of Congress Cataloging-in-Publication Data:

Library of Congress Control Number: 200210096

ISBN: 0-7645-5419-0

Manufactured in the United States of America

10 9 8

3B/RV/QY/QU/IN

About the Author

Howland Blackiston has been a backyard beekeeper for nearly 20 years. He's written many articles on beekeeping and appeared on dozens of television and radio programs (including The Discovery Channel, CNBC, CNN, NPR, and scores of regional shows). He has been a keynote speaker at conferences in more than 40 countries. Howland is cofounder and president of bee-commerce.com, an internet-based super store offering beekeeping supplies and equipment for the hobbyist beekeeper. The Web site receives more than 6 million hits a year. Howland is the past president of Connecticut's Back Yard Beekeepers Association, one of the nation's largest regional clubs for the hobbyist beekeeper. Howland, his wife Joy and their daughter Brooke live in Weston, Connecticut.

Dedication

This book is lovingly dedicated to my wife Joy, who is the queen bee of my universe. She has always been supportive of my unconventional whims and hobbies (and there are a lot of them) and never once did she make me feel like a dummy for asking her to share our lives with honey bees. I also thank our wonderful daughter Brooke, who like her mother, cheerfully puts up with sticky kitchen floors and millions of buzzing "siblings."

Author's Acknowledgments

I was very fortunate, when I started beekeeping, that I met a masterful bee-keeper who took me under his wing and taught me all that is wonderful about honey bees. Ed Weiss became a valued mentor, a great friend, and ultimately a partner in business. I am deeply appreciative of his friendship and bee-wisdom. Ed served as the technical review editor for this book, and I am most appreciative of the many hours he spent checking my facts to ensure that I had been an attentive student. Thank you Ed.

My good friends Anne and David Mayer played a key role in the creation of this book. Both of them are authors, and both encouraged me to contact Wiley. "You should write a book about beekeeping, and they should publish it," they urged. Well, I did and they did. Thank you Anne and David. I owe you a whacking big jar of honey!

A good how-to book needs great how-to images. Special thanks to John Clayton for the stunning cover image and some of the other close-ups used in the book. Thanks also to Steve McDonald and Dr. Edward Ross who provided most of the stunning macrophotography used in this book. I extend my grati-tude for images (and technical suggestions) provided by Kim Flottum at *Bee Culture* magazine. Image credits also go to The National Honey Board, the U.S. Department of Agriculture, Marco Lazzari, Peter Duncan, Eric Erickson, Reg Wilbanks, Mario Espinola, David Eyre, Swienty Beekeeping Equipment, E. H. Thorne Ltd., Wellmark International, Barry Birkey, and Kate Solomon. And thanks to fellow beekeeper and friend Stephan Grozinger, who patiently served as my model for some of the how-to photographs.

Thanks also to Ellen Zampino for her section on planting flowers for your bees, and to Patty Pulliam for her beeswax recipes.

Writing this book was a labor of love, thanks to the wonderful folks at Wiley: Kira Sexton, my acquisitions editor; Neil Johnson, who did the bulk of the copy editing; Pam Mourouzis, who also did some editing and provided sound feedback; and my enthusiastic project editor Suzanne Snyder, whom I might have convinced to become a beekeeper. I would also like to thank Maridee Ennis, the book's production coordinator, who handled nearly everything to do with the way the words and images ultimately appeared on the page. What a great team!

Publisher's Acknowledgments

We're proud of this book; please send us your comments through our online registration form located at www.dummies.com/register.

Some of the people who helped bring this book to market include the following:

Acquisitions, Editorial, and Media Development

Project Editor: Suzanne Snyder

Acquisitions Editor: Kira Sexton

Copy Editor: Neil Johnson

Technical Editor: Ed Weiss

Editorial Manager: Pam Mourouzis

Editorial Assistant: Carol Strickland

Cover Photo: John Clayton

Compostition

Project Coordinator: Maridee Ennis

Layout and Graphics: Joyce Haughey, Jackie Nicholas, Brent Savage, Jacque Schneider, Mary J. Virgin, Erin Zeltner

Proofreaders: John Greenough, Andy Hollandbeck, Marianne Santy, TECHBOOKS Production Services

Indexer: TECHBOOKS Production Services

Special Art: John Clayton, Steve McDonald, Dr. Edward Ross, Kim Flottum at *Bee Culture* magazine, The National Honey Board, U.S. Department of Agriculture, Peter Duncan, Eric Erickson, Mario Espinola, David Eyre, Swienty Beekeeping Equipment, E. H. Thorne Ltd., Wellmark International, Barry Birkey, and Kate Solomon

Publishing and Editorial for Consumer Dummies

Diane Graves Steele, Vice President and Publisher, Consumer Dummies

Joyce Pepple, Acquisitions Director, Consumer Dummies

Kristin A. Cocks, Product Development Director, Consumer Dummies

Michael Spring, Vice President and Publisher, Travel

Brice Gosnell, Associate Publisher, Travel

Suzanne Jannetta, Editorial Director, Travel

Publishing for Technology Dummies

Richard Swadley, Vice President and Executive Group Publisher

Andy Cummings, Vice President and Publisher

Composition Services

Gerry Fahey, Vice President of Production Services

Debbie Stailey, Director of Composition Services

Contents at a Glance

Cartoons at a Glance

By Rich Tennant

page 5

page 195

"Honey - get the smoker. The guests are starting to swarm all over the bee hives again."

page 41

"I try to see to it that it's not all work for my bees."

page 141

"Remind me the next time we're harvesting honey not to sit down and pet the cat afterwards."

page 223

"Oh really, Gerald! You're removing honey from the hives, not plutonium!"

page 97

Cartoon Information:
Fax: 978-546-7747
E-Mail: richtennant@the5thwave.com
World Wide Web: www.the5thwave.com

Table of Contents

Foreword

I'm a hobby beekeeper. I have a few hives in my backyard, like thousands and thousands of other hobby beekeepers the world over. For me, it's an entertaining, educational, and sometimes profitable hobby. But beekeeping is only a hobby — my days are spent gathering information on . . . what else, honey bees and beekeeping.

Information gathering is the easy part of my job, actually. There's lots of information out there (more, it is said, than for any other animal except man). In fact, there is so much information available — hundreds of books, with new titles appearing every year, professional articles by scientists and researchers published in any of several prestigious journals and popular magazines, Web pages by the score, newsletters . . . the list is long.

Actually, there's too much information out there. Finding it all is a challenge, and sorting it out can be a daunting task, even for those experienced in the craft. Sifting through the books, journals, articles, and Web pages to find the gems is what fills my days. A kernel here and a nugget there, when pieced together, makes each issue of the magazine I edit a valuable tool for readers. And staying just ahead of the newest curves is what keeps this job exciting and worthwhile. I, too, need to know what to do for my bees.

The second most-asked question I encounter in my position is "Where do I start?" (The first is "Have you ever been stung?" To that question my answer is always the same, "Yes, but not often.")

Now, to answer the second most-asked question, I will point out Howland's book. It supplies all the information a beginner needs to start keeping bees with confidence and enthusiasm. And it answers all the questions that I have to answer on a daily basis, including that first most-asked question.

My job has become much easier. And beekeeping will be easier and more rewarding for you if you use the information contained on the pages of this book. And, like those thousands and thousands of other backyard beekeepers, keeping bees will be entertaining, enjoyable, and maybe even profitable.

Kim Flottum
Editor, *Bee Culture Magazine*
President, Eastern Apicultural Society

Introduction

Keeping honey bees is a unique and immensely rewarding hobby. If you have an interest in nature, you'll deeply appreciate the wonderful world that beekeeping opens up to you. If you're a gardener, you'll treasure the extra bounty that pollinating bees bring to your fruits, flowers, and vegetables. In short, you'll be captivated by these remarkable little creatures in the same way others have been captivated for thousands of years.

Becoming a beekeeper is easy and safe — it's a great hobby for the entire family. All you need is a little bit of guidance to get started. And that's exactly what this book is for. I provide you with a step-by-step approach for successful backyard beekeeping — follow it closely, and you can have a lifetime of enjoyment with your bees.

What I Assume about You

If you've never kept bees, this book has all the information you need to get started in beekeeping. I assume that you have no prior knowledge of the equipment, tools, and techniques — complete ignorance, in the best sense of the word!

However, if you've been a beekeeper for a while, this book is a terrific resource for you, too. You'll find new ideas on how to keep your bees healthier and more productive. You may appreciate the way the book has been organized for easy and ongoing reference. I include the latest information on honey bee health and medications, plus a whole lot of "tricks of the trade." In short, this book is for just about anyone who's fallen in love with the bountiful honey bee.

How This Book is Organized

This book is a reference, not a lecture. You certainly don't have to read it from beginning to end unless you want to. I organized the chapters in a logical fashion, with sensitivity to the beekeeper's calendar of events. I include lots of great photographs and illustrations (each, I hope, is worth a thousand words) and lots of practical advice and suggestions. The following sections describe how the book is structured:

Part 1: Falling in Love with a Bug

Before becoming a beekeeper, take a moment to get to know the honey bee.

Chapter 1 explains basic bee anatomy and how bees communicate with each other. It also introduces you to the various kinds of honey bees and other stinging insects.

Chapter 2 gives you some insight into "a day in the life of the honey bee." You find out about the queen, the workers, and the drones, and the roles each plays in the colony.

Part 11: Starting Your Adventure

This is where the fun begins! Here's where you find out how to get started with your first colony of honey bees.

Chapter 3 deals with any apprehensions you may have about beekeeping (stings, neighbors, and so on). This chapter tells you where you should locate your hive and how you can get started.

Chapter 4 shows the basic equipment you need and how to assemble it. You find out about really cool gadgets and weird and wonderful hives.

Chapter 5 helps you decide the kind of honey bee to raise, and when and how to order your bees. Find out what to do the day your "girls" arrive and how to successfully transfer them to their new home.

Part 111: Time for a Peek

Here's where you get up-close and personal with your honey bees. This is the heart of the book because it shares useful tips and techniques that help you develop good habits right from the start. You find out the best and safest way to inspect and enjoy your bees.

Chapter 6 clearly explains how to go about approaching and opening up a colony of bees.

Chapter 7 helps you understand exactly what you're look for every time you inspect a colony. I include the specific tasks that are unique to the weeks immediately following the arrival of your bees.

Chapter 8 discusses the tasks a beekeeper must perform year-round to maintain a healthy colony. Use it as a checklist of seasonal activities that you can refer back to year after year.

Part IV: Common Problems & Simple Solutions

Okay, I admit it. Sometimes things go wrong. But don't worry. This section tells you what to expect and what to do when things don't go as planned.

Chapter 9 shows you how to anticipate a number of the most common problems. Find out what to do if your hive swarms or simply packs up and leaves. Discover how to recognize problems with brood production and your precious queen.

Chapter 10 takes a detailed look at bee illnesses. Learn what medications you can use to keep your bees healthy and productive, year after year.

Chapter 11 shows you how to deal with some common pests of the honey bee — mites, birds, insects, and other troublesome critters.

Part V: Sweet Rewards

This is what beekeeping is all about for most people — the honey harvest!

Chapter 12 gets you ready for your honey harvest. Decide what kind of honey you'd like to make. Find out about the equipment you need and how to plan for the big harvest.

Chapter 13 gives you a step-by-step approach for harvesting, bottling, and marketing your honey. The chapter also includes some practical advice for what to do after the harvest is over.

Part VI: The Part of Tens

No *For Dummies* book is complete without the Part of Tens, so I offer a collection of fun lists. Not a bad way to squeeze a whole bunch of extra, helpful information into a book.

Chapter 14 lists ten fun, bee-related activities, including information about starting an observation hive, brewing honey wine, building your own hives, and making products from beeswax and propolis.

Chapter 15 answers the most frequently asked questions about bee behaviors that I've received from beekeepers.

Chapter 16 includes ten of my all-time favorite honey recipes. After all, there's a lot more uses for honey than just spreading it on toast!

I also include some back-of-book materials, including a lot of really helpful bee-related resources: Web sites, journals, suppliers, and beekeeping associations. I also give you a glossary of bee and beekeeping terms that you can use as a handy quick reference, and some useful templates for creating your own beekeeping checklists and logs. Finally, there are some discount coupons that you can apply toward the purchase of new beekeeping equipment and magazine subscriptions.

Icons Used in This Book

Peppered throughout this book are helpful icons that present special types of information to enhance your reading experience:

Think of these tips as words of wisdom that — when applied — can make your beekeeping experience more pleasant and fulfilling!

These warnings alert you to potential beekeeping boo-boos that could make your experiences unpleasant and/or your little winged friends unhappy or downright miserable. Take them to heart!

I use this icon to point out things that need to be so ingrained in your beekeeping consciousness that they become habits. Keep these points at the forefront of your mind when caring for your bees.

From time to time, I explain some new terminology that is basic beekeeping parlance. Learn some new words and some insights to the world of the hive!

Here I share with you some personal beekeeping anecdotes and "betcha didn't know" facts about these winged wonders!

Part I
Falling in Love with a Bug

In this part . . .

*H*ere's where you get know more about the remarkable honey bee. See what makes them tick, understand how they communicate with each other, and find out about their different roles and responsibilities as members of the colony.

Chapter 1

To Bee or Not to Bee?

I've been keeping bees in my backyard for nearly 20 years, and I have a confession to make — I really love my bees. That may sound weird to you if you aren't a beekeeper (yet!), but virtually everyone who keeps bees will tell you the same thing and speak with deep warmth about "their girls." They impatiently await their next opportunity to visit their hives. They experience a true emotional loss when their bees don't make it through a bad winter. Beekeepers, without a doubt, develop a special bond with their bees.

Since becoming a backyard beekeeper, I've grown to deeply admire the remarkable qualities of these endearing creatures. As a gardener, I've witnessed firsthand the dramatic contribution they provide to plants of all kinds. With honey bees in my garden, its bounty has increased by leaps and bounds. And then there's that wonderful bonus that they generously give me: a yearly harvest of sweet liquid gold.

Once you get to know more about bees' value and remarkable social skills, you'll fall in love with them too. They're simply wonderful little creatures. Interacting with them is an honor and a privilege. People who love nature in its purest form will love bees and beekeeping.

That being said, in this chapter, I help you better understand the remarkable and bountiful little honey bee by looking at its history and the value that it brings to our lives. I also discuss the benefits of beekeeping and why you should consider it as a hobby — or even a small business venture. This chapter gives you an idea of what equipment you'll need to get started, the time you should expect to spend maintaining a healthy hive, and how deep your pockets need to be. It also discusses the optimal environmental conditions for raising bees and ends with a checklist that you can fill out to see if beekeeping is for you.

The prehistoric bee

Bees have been around for a long, long time, gathering nectar and pollinating flowers. They haven't changed much since the time of the dinosaurs. The insect shown in the following figure is definitely recognizable as a bee. It was caught in a flow of pine sap 30 to 40 million years ago and is forever preserved in amber.

Courtesy of Mario Espinola, www.espd.com

Discovering the Benefits of Beekeeping

Why has mankind been so interested in beekeeping over the centuries? I'm sure that the first motivator was *honey.* After all, for many years and long before cane sugar, honey was the primary sweetener in use. I'm also sure that honey remains the principal draw for many backyard beekeepers. Chapters 12 and 13 deal with how to produce, harvest, and market your honey.

But the sweet reward is by no means the only reason folks are attracted to beekeeping. For a long time, agriculture has recognized the value of pollination by bees. Without the bees' help, many commercial crops would suffer serious consequences. Even backyard beekeepers witness dramatic improvements in their gardens' yields: more and larger fruits, flowers and vegetables. A hive or two in the garden makes a big difference in your success as a gardener.

The rewards of beekeeping extend beyond honey and pollination. Bees produce other products that can be harvested and put to good use, including beeswax, propolis, and royal jelly. Even the pollen they bring back to the hive can be harvested (it's rich in protein and makes a healthy food supplement in our own diets). For more information about what you can do with these products, be sure to see Chapter14.

Harvesting liquid gold: Honey

The prospect of harvesting honey is certainly a strong attraction for new bee-keepers. There's something magical about bottling your own honey. And I can assure you that no other honey tastes as good as the honey made by your own bees. Delicious! Be sure to have a look at Chapter 16, where I list some delicious recipes for cooking with honey.

How much honey can you expect? The answer to that question varies depending on the weather, rainfall, and location and strength of your colony. But producing 100 pounds or more of surplus honey isn't unusual for a single colony. Chapters 12 and 13 provide plenty of useful information on the kinds of honey you can harvest from your bees and how to go about it. Also included are some suggestions on how you can go about selling your honey — how many hobbies can boast a profitable return on investment!

Bees as pollinators: Experiencing a more bountiful garden

Any gardener recognizes the value of pollinating insects. Various insects perform an essential service in the production of seed and fruit. The survival of plants depends on pollination, and the honey bee accounts for 80 percent of all pollination done by insects. Without the honey bee's services, more than a third of the fruits and vegetables that humans consume would be lost.

Honeybee or honey bee?

This is a "tomato/to*mah*to" issue. The British adhere to their use of the one word: "honey-bee." The Entomological Society of America, however, prefers to use 2 words "honey bee." Here's the Society's rationale: The honey bee is a true bee, like a house fly is a true fly, and thus should be 2 words. A dragonfly, on the other hand, is not a fly; hence it is one word. **Note:** Spell it both ways when Web surfing. That way, you'll cover all bases and hit all the sites!

Why bees make great pollinators

About 90 crops in the United States depend on bees for pollination. Why is the honey bee such an effective pollinator? Because she's uniquely adapted to the task. Here are several examples:

- The honey bee's anatomy is well suited for carrying pollen. Her body and legs are covered with branched hairs that catch and hold pollen grains. The bee's hind legs contain *pollen baskets* (see the photo on "The Bountiful Bee" page in the color section of this book) that the bee uses for transporting pollen, a major source of food, back to the hive. If the bee brushes against the stigma (female part) of the next flower she visits and brushes off some of the pollen grains, the act of cross-pollination is accomplished.

- Most other insects lie dormant all winter and in spring emerge only in small numbers, until increasing generations have rebuilt the population of the species. Not the honey bee. Its hive is perennial. The honey bee overwinters, with large numbers of bees feeding on stored honey. Early in the spring, the queen begins laying eggs and the already large population explodes. When flowers begin to bloom, each hive has tens of thousands of bees to carry out pollination activities. By mid-summer, an individual hive contains upward of 60,000 bees.

- The honey bee has a unique habit that's of great value as a pollinator. It tends to forage on blooms of the same kind, as long as they're flowering. In other words, rather than hopping from one flower type to another, honey bees are flower-consistent. This focus makes for particularly effective pollination. It also means that the honey they produce from the nectar of a specific flower takes on the unique flavor characteristics of that flower — that's how we get specific honey flavors, such as orange blossom honey, buckwheat honey, blueberry honey, lavender honey, and so on (see Chapter 3).

- The honey bee is one of the only pollinating insects that can be introduced to a garden at the gardener's will. You can garden on a hit-or-miss basis and hope that enough wild bees are out there to achieve adequate pollination — or you can take positive steps and nestle a colony of honey bees in a corner of your garden.

I've witnessed the miracle in my own garden: more and bigger flowers, fruits, and vegetables — all the result of more efficient pollination by bees. After seeing my results, a neighbor who tends an imposing vegetable garden begged me to place a couple of hives on her property. I did, and she too is thrilled. She rewards me with a never-ending bounty of fruits and vegetables. And I pay my land-rent by providing her with 20 pounds of honey every year. Not a bad barter all around.

Being part of the bigger picture: Save the bees!

The facts that keeping a hive in the backyard dramatically improves pollination and rewards you with a delicious honey harvest are by themselves good enough reasons to keep bees. But today, the value of keeping bees goes beyond the obvious. In many areas, millions of colonies of wild (or *feral*) honey bees have been wiped out by urbanization, pesticides, and parasitic mites, devastating the wild honey bee population. Many gardeners have asked me why they now see fewer and fewer honey bees in their gardens. It's because of the dramatic decrease in our wild honey bee population. Backyard beekeeping has become vital in our efforts to reestablish lost colonies of bees and offset the natural decrease in pollination by wild bees.

Bee hunters, gatherers, and cultivators

An early cave painting in eastern Spain, circa 6000 b.c., shows early Spaniards hunting for and harvesting wild honey (see the figure below). In centuries past, honey was a treasured and sacred commodity. It was used as money and praised as the nectar of the gods. Methods of beekeeping remained relatively unchanged until 1852 with the introduction of today's "modern" interchangeable-frame hive, also known as the Langstroth hive. (See Chapter 4 for more information about Langstroth and other kinds of bee hives.)

Bee pollen, honey, and allergy relief

Pollen is one of the richest and purest of natural foods, consisting of up to 35 percent protein and 10 percent sugars, carbohydrates, enzymes, minerals, and vitamins A (carotenes), B1 (thiamin), B2 (riboflavin), B3 (nicotinic acid), B5 (panothenic acid), C (ascorbic acid), H (biotin), and R (rutine).

Here's the really neat part: Ingesting small amounts of pollen every day can actually help reduce the symptoms of pollen-related allergies — sort of a homeopathic way of inoculating yourself.

Of course you can harvest pollen from your bees, and sprinkle a small amount on your breakfast cereal or in yogurt (as you might do with wheat germ). But you don't really need to harvest the pollen itself. That's because raw, natural honey contains pollen. Pollen's benefits are realized every time you take a tablespoon of honey. Eating local honey every day can relieve the symptoms of pollen-related allergies, if the honey is harvested from within a 50-mile radius of where you live or from an area where the vegetation is similar to what grows in your community. Now that you have your own bees, that isn't a problem. Allergy relief is only a sweet tablespoon away!

Getting an education: And passing it on!

As a beekeeper you continually discover new things about nature, bees, and their remarkable social behavior. Just about any school, nature center, garden club, or youth organization loves for you (as a beekeeper) to share your knowledge. Each year I make the rounds with my slide show and props, sharing the miracle of honey bees with my community. On many occasions my daughter's teacher and classmates visited the house for an on-site workshop. I opened the hive and gave each wide-eyed student a close-up look at bees at work. Spreading the word to others about the value these little creatures bring to all of us is great fun. You're planting a seed for our next generation of beekeepers. After all, a grade-school presentation on beekeeping is what aroused my interest in honey bees.

Improving your health: Bee therapies and stress relief

Although I can't point to any scientific studies to confirm it, I honestly believe that tending honey bees reduces stress. Working with my bees is so calming and almost magical. I am at one with nature, and whatever problems may have been on my mind tend to evaporate. There's something about being out there on a lovely warm day, the intense focus of exploring the wonders of the hive, and hearing that gentle hum of contented bees — it instantly puts me at ease, melting away whatever day-to-day stresses that I might find creeping into my life.

Any health food store proprietor can tell you the benefits of the bees' products. Honey, pollen, royal jelly, and propolis have been a part of healthful remedies for centuries. Honey and propolis have significant antibacterial qualities. Royal jelly is loaded with B vitamins and is widely used overseas as a dietary and fertility stimulant. Pollen is high in high protein and can be used as a homeopathic remedy for seasonal pollen allergies (see "Bee pollen, honey, and allergy relief" sidebar earlier in this chapter).

Apitherapy is the use of bee products for treating health disorders. Even the bees' venom plays an important role here — in bee-sting therapy. Venom is administered with success to patients who suffer from arthritis and other inflammatory/medical conditions. This entire area has become a science in itself and has been practiced for thousands of years in Asia, Africa, and Europe. An interesting book on apitherapy is *Bee Products — Properties, Applications and Apitherapy: Proceedings of an International Conference Held in Tel Aviv, Israel, May 26–30, 1996*, published by Kluwer Academic Publishers (ISBN: 0306455021).

More information on apitherapy is available from the American Apicultural Society (www.apitherapy.org).

Determining Your Beekeeping Potential

How do you know whether you'd make a good beekeeper? Is beekeeping the right hobby for you? Here are a few things worth considering as you ponder these issues.

Environmental considerations

Unless you live on a glacier or on the frozen tundra of Siberia, you probably can keep bees. Bees are remarkable creatures that do just fine in a wide range of climates. Beekeepers can be found in areas with long cold winters, in tropical rain forests, and in nearly every geographic region in-between. If flowers bloom in your part of the world, you can keep bees.

How about space requirements? You don't need much. I know many beekeepers in the heart of Manhattan. They have a hive or two on their rooftops or terraces. Keep in mind that bees travel miles from the hive to gather pollen and nectar. They'll forage an area as large as 6,000 acres, doing their thing. So the only space that you need is enough to accommodate the hive itself.

Knowing where honey bees come from

Honey bees are native to Europe, Asia, and Africa, but they're not native to other parts of the world (Australia, New Zealand, and the Americas). Colonies of honey bees were first shipped to Virginia in 1621 and their honey was used by the early pioneers as their chief sweetener. These bees prospered and gradually colonized all of North America. Today, they've become a vital part of our agricultural economy. Honey bees didn't reach Australia and New Zealand until the early to mid-1800s.

See Chapter 3 for more specific information on where to locate your bees.

Zoning and legal restrictions

Most communities are quite tolerant of beekeepers, but some have local ordinances that prohibit beekeeping or restrict the number of hives that you can have. Some communities let you keep bees but ask that you register your hives with the local government. Check with your town hall, local zoning board, or state agricultural experiment station to find out about what's okay in your neighborhood.

Obviously you want to practice a good-neighbor policy, so that folks in your community don't feel threatened by your unique new hobby. See Chapter 3 for more information on the kinds of things you can do to prevent neighbors from getting nervous.

Costs and equipment

What does it cost to become a beekeeper? All in all, beekeeping isn't a very expensive hobby. You can figure on investing about $200 to $400 for the hive, equipment, tools, and medication. In addition, you'll spend $50 to $70 for a package of bees and queen. For the most part, these are one-time expenses. Keep in mind, however, the potential for a return on this investment. Your hive can give you 100 to 200 pounds of honey every year. At $4 a pound (a fair going price for all-natural, raw honey), that should give you an income of $400 to $800 per hive! Not bad, huh?

See Chapter 4 for a detailed listing of the equipment you'll need.

How many hives do you need?

Most beekeepers start out with one hive. And that's probably a good way to start your first season. But most beekeepers wind up getting a second hive in short order. Why? For one, it's twice as much fun! Another more practical reason for having a second hive is that recognizing normal and abnormal situations is easier when you have two colonies to compare. In addition a second hive enables you to borrow frames from a stronger, larger colony to supplement one that needs a little help. My advice? Start with one hive until you get the hang of things, and then consider expanding in your second season.

What kind of honey bees should you raise?

The honey bee most frequently raised by beekeepers in the United States today is European in origin and has the scientific name *Apis mellifera*.

Of this species, the most popular bee is the so-called "Italian" honey bee. These bees are docile, hearty, and good honey producers. They are a good choice for the new beekeeper. See Chapter 5 for more information about different varieties of honey bees.

Time and commitment

Beekeeping isn't labor intensive. Sure you'll spend a weekend putting together your new equipment. And I'm anticipating that you'll be spending some time reading up on your new hobby. (I sure hope you read my book from cover to cover!) But the actual time that you absolutely *must* spend with your bees is surprisingly modest. Other than your first year (when I urge you to inspect the hive frequently to find out more about your bees) you need to make only five to six visits to your hives every year. Add to that the time that you spend harvesting honey, repairing equipment, and putting things away for the season, and you'll probably devote 19 to 28 hours a year to your hobby (more if you make a business out of it).

For a more detailed listing of seasonal activities, be sure to read Chapter 8.

Beekeeper personality traits

If you run like a banshee every time you see an insect, I suspect that beekeeping will be an uphill challenge for you. But if you love animals, nature, and the outdoors, and if you're curious about how creatures communicate and contribute to our environment, you'll be captivated by honey bees. If you like the idea of "farming" on a small scale, or you're intrigued by the prospect of harvesting your own all-natural honey, you'll enjoy becoming a beekeeper. Sure, as far as hobbies go, it's a little unusual, but all that's part of its allure. Express your uniqueness and join the ranks of some of the most delightful and interesting people I've ever met . . . backyard beekeepers!

Honey trivia

There are all kinds of interesting facts about honey. Here's a hodgepodge of trivia that might improve your chances of winning a quiz show.

- Honey has antibacterial properties and is used in some cultures to prevent infection of cuts and burns. A medico friend of mine recently visited a burn clinic in China where honey is used in the patients' dressings.

- In olden days, a common practice was for newlyweds to drink mead (honey wine) for one month (one phase of the moon) to assure the birth of a son. Thus the term "honeymoon."

- The honey bee's image became a symbol for kings and religious leaders and was honored on ancient coins and in mythology.

- One gallon of honey (3.79 liters) weighs 11 lbs., 13.2 ounces (5.36 kg.).

- The Romans used honey to pay their taxes (I don't think the IRS would approve).

- Honey found in the tombs of the Egyptian Pharaohs was still edible. That's an impressive shelf life!

Allergies

If you're going to become a beekeeper, you can expect to get stung once in a while. It's a fact of life. But when you learn good habits as a beekeeper, you can minimize or even eliminate the chances that you'll be stung.

All bee stings can hurt a little, but not for long. It's natural to experience some swelling, itching, and redness. These are *normal* (not allergic) reactions. Some folks are mildly allergic to bee stings, and the swelling and discomfort may be more severe. And yet, the most severe and life-threatening reactions to bee stings occur in less than 1 percent of the population. So the chances that you're dangerously allergic to honey bee venom are remote. If you're uncertain, check with an allergist, who can determine whether you're among the relatively few who should steer clear of beekeeping.

You'll find more information on bee stings in Chapter 3.

Chapter 2

Life Inside the Honey Bee Hive

My first introduction to life inside the honey bee hive occurred many years ago during a school assembly. My classmates and I were shown a wonderful movie about the secret inner workings of the beehive. The film mesmerized me. I'd never seen anything so remarkable and fascinating. How could a bug be so smart and industrious? I couldn't help being captivated by the bountiful honey bee. That brief childhood event planted a seed that blossomed into a treasured hobby some 20 years later.

Anyone who learns even a little bit about the honey bee can't help but be amazed, because far more goes on within the hive than most people can ever imagine: complex communication, social interactions, teamwork, unique jobs and responsibilities, food gathering, and the engineering of one of the most impressive living quarters found in nature. Whether newcomer or old hand, you'll have many opportunities to experience first-hand the miracle of beekeeping. Every time that you visit your bees you see something new. But you'll get far more out of your new hobby if you understand more about what you're looking at. What are the physical components of the bee that enable it to do its job so effectively? What are those bees up to and why? What's normal and what's not normal? What is a honey bee and what is an imposter? In this chapter you'll take a peek within a typical colony of honey bees.

Basic Body Parts

Everyone knows about at least one part of the honey bee's anatomy: its stinger. But you'll get more out of beekeeping if you understand a little bit about the other various parts that make up the honey bee. I won't go into this in textbook detail — just a few basic parts (see Figure 2-1) to help you understand what makes them tick.

Skeleton

Like all insects, the honey bee's "skeleton" is on the outside. This arrangement is called an *exoskeleton*. Nearly the entire bee is covered with *branched hairs* (like the needles on the branch of a spruce tree). A bee can "feel" with these hairs, and the hairs serve the bee well when it comes to pollination, because pollen sticks well to the branched hairs.

Head

The honey bee's head (see Figure 2-2) is flat and somewhat triangular in shape. Here's where you'll find the bee's brain and primary sensory organs (sight, feel, taste, and smell). It's also where you'll find important glands that produce royal jelly and various chemical pheromones used for communication.

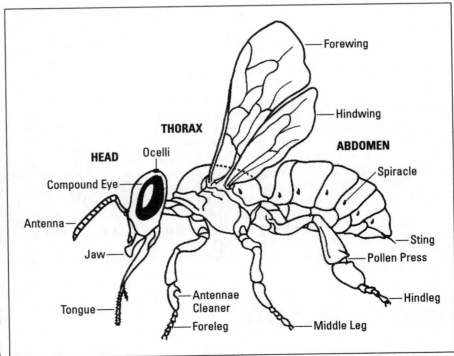

Figure 2-1: This is how a honey bee looks if you shave all the hairs off. The illustration shows the basic body parts of the bee.

Royal jelly is a substance secreted from glands in a worker bee's head and used as a food to feed brood.

The important parts of the bee's head are its:

- **Eyes.** The head includes two large compound eyes that are used for general-distance sight and three small simple eyes, called ocelli, which are used in the poor light conditions within the hive. Notice the three simple eyes (ocellus) on the members of all three castes in Figure 2-2, while the huge wrap-around compound eyes of the drone make him easy to identify. The queen's eyes, however, are slightly smaller than the worker bee.

- **Antennae.** The honey bee has two antennae in front (attached to its forehead). Each antenna has thousands of tiny sensors that detect smell (like a nose does). The bee uses this sense of smell to identify flowers, water, the colony, and maybe even you! They also, like the branched hairs mentioned earlier, detect feel.

- **Mouth parts.** The bees' mandibles (jaws) are used for feeding larvae, collecting pollen, manipulating wax, and carrying things.

- **Proboscis.** Everyone's familiar with those noisemakers that show up at birthday and New Year's Eve parties. You know, the ones that unroll when you toot them! The bee's proboscis is much like those party favors only without the "toot." When the bee is at rest, this organ in retracted. But when the bee is feeding or drinking, it unfolds to form a long tube that the bee uses like a straw.

Figure 2-2:
Comparing the heads of worker, drone, and queen bees, note the worker bee's extra-long proboscis and the drone's huge wraparound eyes.

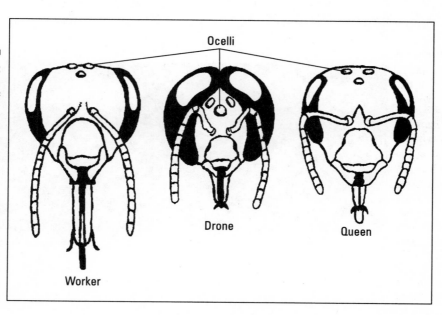

Thorax

The thorax composes the middle part of the bee. It is the segment between the head and the abdomen where the two pairs of wings and six legs are anchored.

- **Wings.** Here's a question for you: How many wings does a honey bee have? The answer is four. Two pairs are attached fore and aft to the bee's thorax. The wings are hooked together in flight and separate when the bee is at rest.

- **Legs.** The bee's three pairs of legs all are different. Each leg has six segments that make them quite flexible. The bees also have taste receptors on the tips of their legs. The bee uses its forward-most legs to clean its antennae. The middle legs help with walking and are used to pack loads of pollen (and sometimes propolis) onto the *pollen baskets* that are part of the hind legs. *Propolis* is the sticky resinous substance that the bees collect from plants and use to seal up cracks in the hive. Propolis can be harvested and used for a variety of nifty products. For more information on propolis and what you can do with it, see Chapter 14. The hind legs (see Figure 2-3) are specialized on the worker bee. They contain special combs and a pollen press, which are used by the worker bee to brush, collect, pack, and carry pollen and propolis back to the hive. Take a moment to watch a foraging bee on a flower. You'll see her hind legs heavily loaded with pollen for the return trip home.

- **Spiracles.** These tiny holes along the bee's thorax and abdomen are the means by which a bee breathes. The bee's trachea (breathing tubes) are attached to these spiracles. It is through these holes that tracheal mites gain access to the trachea.

Figure 2-3:
In this close-up image of a bee's leg, you can clearly see the hairs that serve as brushes to collect pollen.

Courtesy of Dr. Eric Erickson, Jr.

Abdomen

The abdomen is the part of the bee's body that contains its digestive organs, reproductive organs, wax and scent glands (workers only), and, of course, the infamous stinger (workers and queen only).

The Amazing Language of Bees

It is said that only man has a form of communication superior to that of the honey bee. Like you and I, honey bees utilize five senses throughout their daily lives; however, honey bees have additional communication aids at their disposal. Two of the methods by which they communicate are of particular interest. One is chemical, the other choreographic.

Pheromones

What are *pheromones*? They're chemical scents that animals produce to trigger behavioral responses from the other members of the same species. Honeybee pheromones provide the "glue" that holds the colony together. The three castes of bees, of which more is mentioned later in this chapter, produce various pheromones at various times to stimulate specific behaviors. The study of pheromones is a topic worthy of an entire book, so here are just a few basic facts about the ways pheromones help bees communicate:

✔ Certain queen pheromones (known as *queen substance,* discussed at greater length later in this chapter) let the entire colony know that the queen is in residence and stimulate many worker bee activities.

✔ Outside of the hive, the queen pheromones act as a sex attractant to potential suitors (male drone bees). They also regulate the drone (male bee) population in the hive.

✔ Queen pheromones stimulate many worker bee activities, such as comb building, brood rearing, foraging, and food storage.

✔ The worker bees at the hive's entrance produce pheromones that help guide foraging bees back to their hive. The Nassanoff gland (discussed later in this chapter) at the tip of the worker bee's abdomen is responsible for this alluring scent.

✔ Worker bees produce alarm pheromones that can trigger sudden and decisive aggression from the colony.

✔ The colony's brood (developing bee larvae and pupae) secretes special pheromones that help worker bees recognize the brood's gender, stage of development, and feeding needs.

Shall we dance?

Perhaps the most famous and fascinating "language" of the honey bee is communicated through a series of dances done by foraging worker bees who return to the hive with news of nectar, pollen, or water. The worker bees dance on the comb using precise patterns. Depending upon the style of dance, a variety of information is shared with the honey bees' sisters. They're able to obtain remarkably accurate information about the location and type of food the foraging bees have discovered.

Two common types of dances are the so-called *round dance* and the *waggle dance*. The round dance communicates that the food source is near the hive (within 10–80 yards).

For a food source found at a greater distance from the hive, the worker bee performs the waggle dance. It involves a shivering side-to-side motion of the abdomen, while the dancing bee forms a figure eight (see Figure 2-4). The vigor of the waggle, the number of times it is repeated, the direction of the dance, and the sound the bee makes communicates amazingly precise information about the location of the food source.

Figure 2-4:
The worker
bee in the
center of
this image
performs a
dance that
communi-
cates a
specific
food source
and
location.

Courtesy of Dr. Edward Ross, California Academy of Sciences

The dancing bees pause between performances to offer potential recruits a taste of the goodies they bring back to the hive. Combined with the dancing, the samples provide additional information about where the food can be found and what type of flower it is from.

Dividing Honey Bees into Three Castes

During summer months, about 60,000 or more bees reside in a healthy hive. And while you may think that all those insects look exactly alike, actually three different castes (worker, queen, and drone) make up the total population. Each has its own characteristics, roles, and responsibilities. Upon closer examination, the three types even look a little different, and being able to distinguish one from the other is important.

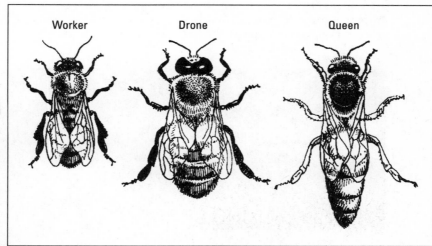

Worker	Drone	Queen

Figure 2-5: These are the three castes of honey bee: worker, drone, and queen.

Her majesty, the queen

Let there be no mistake about it — the queen bee is the heart and soul of the colony. She is the reason for nearly everything the rest of the colony does. The queen is the only bee without which the rest of the colony cannot survive. Without her, your hive is sunk. A good quality queen means a strong and productive hive. For more information on how to evaluate a good queen, see Chapter 7.

As a beekeeper, on every visit to the hive you'll need to determine "do I have a queen?" and "is she healthy?"

Only one queen lives in a given hive. She is the largest bee in the colony, with a long and graceful body. She is the only female with fully developed ovaries. The queen's two primary purposes are to produce chemical scents that help regulate the unity of the colony and to lay eggs — and lots of them. She is, in fact, an egg-laying machine, capable of producing more than 1,500 eggs a day at 30-second intervals. That many eggs are more than her body weight!

The other bees pay close attention to the queen, tending to her every need. Like a regal celebrity, she's always surrounded by a flock of attendants as she moves about the hive (see Figure 2-6). Yet, she isn't spoiled. These attendants are vital, because the queen is totally incapable of tending to her own basic needs. She can neither feed nor groom herself. She can't even leave the hive to relieve herself. And so her doting attendants (the queen's court) take care of her basic needs while she tirelessly goes from cell to cell doing what she does best . . . laying eggs.

The gentle queen bee has a stinger, but it is rare for a beekeeper to be stung by a queen bee. I have handled many queen bees and have never been stung by any of them. In general, queen bees use their stingers only to kill rival queens that may emerge or be introduced in the hive.

Figure 2-6:
A queen and her attentive attendants.

Courtesy of John Clayton

Amazing "queen substances"

In addition to laying eggs, the queen plays a vital role in maintaining the colony's cohesiveness and stability. The mere presence of the queen in the hive motivates the productivity of the colony. Her importance to the hive is evident in the amount of attention paid to her by the worker bees everywhere she goes in the hive. But, as is true of every working mom or regal presence, she can't be everywhere at once, and she doesn't interact with every member of the colony every day. So, how does the colony know they have a queen? By her scent. The queen produces a number of different pheromones (mentioned earlier in this chapter) in her mandibular (jaw) glands that attract workers to her and stimulate brood rearing, foraging, comb building, and other activities. Also referred to as *queen substances,* these pheromones play an important role in controlling the behavior of the colony: Queen substances inhibit the worker bees from making a new queen and prevent the development of the worker bees' ovaries, thus ensuring that the queen is the only egg-laying female in the hive. They act as a chemical communication that "all is well — the queen is in residence and at work." As a queen ages, these pheromones diminish, and, when that happens, the colony knows that it's time to supersede her with a new, young queen.

Pheromones are essential in controlling the well being of the colony. This queen substance makes its way around the hive like a bucket brigade. The queen's attendants pick up the scent from the queen and transfer it by contact with neighboring bees. They in turn pass the scent onto others, and so it distributes throughout the colony. So effective is this relay, that if the queen were removed from the hive, the entire colony would be aware of her loss within hours. When the workers sense the lack of a queen, they become listless and their drive to be productive is lost. Without leadership, they nearly lose their reason for being! First they're unhappy and mope around, but then it dawns on them . . . "let's make a new queen."

The queen can live for two or more years, but replacing your queen after a couple of seasons ensures maximum productivity. Some beekeepers routinely replace their queens every autumn. That practice ensures that your hive has a new energetic young queen each spring. You may wonder why you should replace the queen if she's still alive? That's an easy one: As a queen ages, her egg-laying capability slows down, which results in less and less brood each season. Less brood means a smaller colony. And a smaller colony means a lackluster honey harvest for you! For information on how to successfully introduce a new queen, see Chapter 9.

As a beekeeper, your job is to *anticipate* problems before they happen. An aging queen is something that you can deal with by systematically replacing her before you have a problem.

The industrious little worker bee

The majority of the hive's population consists of *worker bees.* Like the queen, worker bees are all female, but these girls lack fully developed ovaries. They also look different than the queen. They are smaller, their abdomens are shorter, and on their hind legs they possess *pollen baskets,* which are used to tote pollen back from the field.

Like the queen, the worker bee has a stinger. But her stinger is not a smooth syringe like the queen's. It has a barb on the end. The barb causes the stinger, venom sack and a large part of the bee's gut to remain in a human victim — a Kamikaze effort to protect the colony. Only in mammals (such as humans) does the bee's stinger get stuck. The bee can sting other insects again and again while defending its home.

The life span of worker bee is a modest six weeks during the colony's active season. However, worker bees live longer (four to eight months) during the less active winter months.

The term "busy as a bee" is well earned. Worker bees do a considerable amount of work. They do it tirelessly, day in and day out. In fact, during the busy season they literally work themselves to death. The specific jobs and duties they perform during their short lives vary as they age. Understanding their roles will deepen your fascination and appreciation of these remarkable creatures.

From the moment a worker bee is hatched, she has many and varied tasks clearly cut out for her. As she ages, she performs more and more complex and demanding tasks. Although these various duties usually follow a set pattern and timeline, they sometimes overlap. A worker bee may change occupations sometimes within minutes, if there is an urgent need within the colony for a particular task. They represent teamwork and empowerment at their best!

Initially, a worker's responsibilities include various tasks within the hive. At this stage of development, worker bees are referred to as *house bees.* As they get older, their duties involve work outside of the hive as *field bees.*

In the following paragraphs, I highlight the various responsibilities of worker bees during their short but remarkable lives.

Housekeeping (days 1 to 3)

A worker bee is born with the munchies. Immediately after she hatches and grooms herself, she engorges herself with pollen and honey. Following this binge, one of her first tasks is cleaning out the cell from which she just hatched. This and other empty cells are cleaned and polished and left immaculate to receive new eggs and to store nectar and pollen.

Royal jelly: The food of royalty

Royal jelly is the powerful, creamy substance that transforms an ordinary worker bee into a queen bee, and extends her life span from six weeks to five years! It's made of digested pollen and honey or nectar mixed with a chemical secreted from a gland in a nurse bee's head. In health food stores it commands premium prices rivaling imported caviar. Those in the know use royal jelly as a dietary supplement and fertility stimulant. It contains an abundance of nutrients, including essential minerals, B-complex vitamins, proteins, amino acids, collagen, essential fatty acids, just to name a few! Sarah Ferguson ("Fergie"), the former Duchess of York is said to have eaten royal jelly while she was trying to become pregnant.

Undertaking (days 3 to 16)

The honey bee hive is one of the cleanest and most sterile environments found in nature. Preventing disease is an important early task for the worker bee. During the first couple weeks of her life, the worker bee removes any bees that have died and disposes of the corpses as far from the hive as possible. Similarly, diseased or dead brood are quickly removed before becoming a health threat to the colony. Should a larger invader (such as a mouse) be stung to death within the hive, the workers have an effective way of dealing with that situation. Obviously a dead mouse is too big for the bees to carry off. So the workers completely encase the corpse with *propolis* (a brown sticky resin collected from trees, and sometimes referred to as *bee glue*). Propolis has significant antibacterial qualities. In the hot, dry air of the hive, the hermetically sealed corpse becomes mummified and is no longer a source of infection. The bees also use propolis to seal cracks and varnish the inside walls of the hive.

Working in the nursery (days 4 to 12)

The young worker bees tend to their "baby sisters" by feeding and caring for the developing larvae. On average, nurse bees check a single larva 1,300 times a day. They feed the larvae a mixture of pollen and honey, and royal jelly — rich in protein and vitamins — produced from the hypopharyngeal gland in the worker bee's head. The number of days spent tending brood depends upon the quantity of brood in the hive, and the urgency of other competing tasks.

Attending royalty (days 7 to 12)

Because her royal highness is unable to tend to her most basic needs by herself, some of the workers do these tasks for her. They groom and feed the queen, and even remove her excrement from the hive. These royal attendants also coax the queen to continuously lay eggs as she moves about the hive.

Going to the grocery (days 12 to 18)

Young worker bees also take nectar from foraging field bees that are returning to the hive. The house bees deposit this nectar into cells earmarked for this purpose. They add an enzyme to the nectar and set about fanning the cells to evaporate the water content and turn the nectar into ripened honey. The workers similarly take pollen from returning field bees and pack the pollen into cells. Both the ripened honey and the pollen are food for the colony.

Fanning (days 12 to 18)

Worker bees also take a turn at controlling the temperature and humidity of the hive. During warm weather and during the honey flow season, you'll see groups of bees lined up at one side of the entrance, facing the hive. They fan furiously to draw air into the hive. Additional fanners are in position within the hives. This relay of fresh air helps maintain a constant temperature (93 to 95 degrees F) for developing brood. The fanning also hastens the evaporation of excess moisture from the curing honey.

The workers also perform another kind of fanning, but it isn't related to climate control. It has more to do with communication. The bees have a scent gland located at the end of their abdomen called the *Nassanoff gland.* You'll see worker bees at the entrance with their abdomens arched, and the moist pink membrane of this gland exposed (see Figure 2-7). They fan their wings to release this pleasant sweet odor into the air. The pheromone is highly attractive and stimulating to other bees, and serves as an orientation message to returning foragers, saying: "Come hither . . . this is your hive and where you belong."

Figure 2-7:
This worker bee fans her wings while exposing her Nassanoff gland to release a sweet orientation scent. This helps direct other members of the colony back to the hive.

Courtesy of Bee Culture Magazine

Becoming architects and master builders (days 12 to 35)

Worker bees that are about 12 days old are mature enough to begin producing beeswax. These white flakes of wax are secreted from wax glands on the underside of the worker bee's abdomen. They help with the building of new *wax comb* and in the capping of ripened honey and cells containing developing pupae.

Some new beekeepers are alarmed when they first see these wax flakes on the bee. They wrongly think these white chips are an indication of a problem (disease or mite).

Guarding the hive (days 18 to 21)

The last task of a house bee before she ventures out is that of guarding the hive. At this stage of maturity, her sting glands have developed to contain an authoritative amount of venom. You can easily spot the guard bees at the hive's entrance (see photo of guard bees in color insert). They are poised and alert, checking each bee that returns to the hive for a familiar scent. Only family members are allowed to pass. Strange bees, wasps, hornets, and others intent on robbing the hives vast stores of honey are bravely driven off.

Bees from other hives are occasionally allowed in when they bribe the guards with nectar. These bees simply steal a little honey or pollen and leave.

Steppin' out (days 22 to 42)

With her life half over, the worker bee now ventures outside of the hive and joins the ranks of field bees. You'll see them taking their first *orientation flights*. The bees face the hive and dart up, down, and all around the entrance. They're imprinting the look and location of their home before beginning to circle the hive and progressively widening those circles, learning landmarks that ultimately will guide them back home. At this point, worker bees are foraging for pollen (see Figure 2-8), nectar, water, and propolis (resin collected from trees).

Foraging bees visit 5 million flowers to produce a single pint of honey. They forage a two- to three-mile (four- to five-kilometer) radius from the hive in search of food. That's the equivalent of nearly 6,000 acres! So don't think for a moment that you need to provide everything they need on your property. They're ready and willing to travel.

Foraging is the toughest time for the worker bee. It's difficult and dangerous work, and it takes its toll. They can get chilled as dusk approaches and die before they can return to the hive. Sometimes they become a tasty meal for a bird or other insect. You can spot the old girls returning to the hive. They've grown darker in color, and their wings are torn and tattered. This is how the worker bee's life draws to a close . . . working diligently right until the end.

Figure 2-8:
This bee's pollen baskets are filled. She can visit 10 flowers every minute, and may visit more than 600 flowers before returning to the hive.

Courtesy of Wellmark International

The woeful drone

This brings us to the only male bee in the colony. Drones make up a relatively small percentage of the hive's total population. At the peak of the season their numbers may be only in the hundreds. You rarely find more than a thousand.

New beekeepers often mistake a drone for the queen, because he is larger and stouter than a worker bee. But his shape is in fact more like a barrel (the queen's shape is thinner, more delicate and tapered). The drone's eyes are huge and seem to cover his entire head. He doesn't forage for food from flowers — he has no pollen baskets. He doesn't help with the building of comb — he has no wax producing glands. Nor can he help defend the hive — he has no stinger and can be handled by the beekeeper with absolute confidence.

The drone gets a bad rap in many bee books. Described as lazy, glutinous, and incapable of caring for himself, you might even begin wondering what he's good for?

He mates! Procreation is the drone's primary purpose in life. Despite their high maintenance (they must be fed and cared for by the worker bees), drones are tolerated and allowed to remain in the hive because they may be needed to mate with a new virgin queen (when the old queen dies or needs to be superseded). Mating occurs outside of the hive, in mid-flight, 200 to 300 feet in the air. The drone's big eyes come in handy for spotting virgin queens taking their *nuptial flights*. The few drones that do get a chance to mate are in for a sobering surprise. They die after mating! That's because their sex organ is barbed (like the worker bee's stinger). After mating with the queen, their most personal apparatus and a significant part of their internal anatomy is torn away, and they fall to their deaths, a fact that prompts empathetic groans from the men in my lectures and unsympathetic cheers from the women.

Once the weather gets cooler and the mating season comes to a close, the workers will not tolerate having drones around. After all, those fellows have big appetites and would consume a tremendous amount of food during the perilous winter months. So at the end of the nectar-producing season, you will see the worker bees systematically expelling the drones from the hive (see the photo in this book's color section). They are literally tossed out the door. This is your signal that the beekeeping season is over for the year.

The Honey Bee Life Cycle

In winter the hive is virtually dormant. The adult bees are in a tight cluster for warmth, and their queen is snugly safe in the center of it all. But as the days lengthen and the spring season approaches, the bees begin feeding the queen royal jelly. This special food (secreted from the glands near the workers' mandibles) is rich in protein and stimulates the queen to start laying eggs.

Like butterflies, honey bees develop in four distinct phases: egg, larva, pupa, and adult. The total development time varies a bit among the three castes of bees, but the basic miraculous process is the same: 24 days for drones, 21 days for worker bees, and 16 days for queens.

Egg

The metamorphosis begins when the queen lays an egg. You should learn how to spot eggs, because that is one of the most basic and important skills you need to develop as a beekeeper. It isn't an easy task, because the eggs are mighty tiny (only about 1.7 millemeters long). But finding eggs is one of the surest ways to confirm that your queen is alive and well. It's a skill you'll use just about every time you visit your hive.

The queen lays a single egg in each cell that has been cleaned and prepared by the workers to raise new brood (see Figure 2-9). The cell must be spotless, or she moves on to another one.

If she chooses a standard worker-size cell, she releases a fertilized egg into the cell. That egg develops into a worker bee (female). But if she chooses a wider drone-size cell, the queen releases a nonfertilized egg. That egg develops into a drone bee (male). The workers that build the cells are the ones that regulate the ratio of female worker bees to male drone bees. They do this by building smaller cells for female worker bees, and larger cells for male drone bees.

Figure 2-9:
Note the rice-like shape of the eggs and how the queen has positioned them "standing up" in the cell.

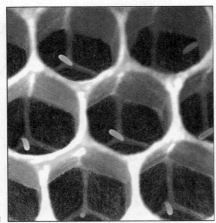

Courtesy of Stephen McDaniel

The queen positions the egg in an upright position (standing on end) at the bottom of a cell. That's why they're so hard to see. When you look straight down into the cell, you're looking at the miniscule diameter of the egg, which is only 0.4 of a millimeter wide. Figure 2-10 shows a microscopic closeup of a single egg.

Eggs are much easier to spot on a bright sunny day. Hold the comb at a slight angle, and with the sun behind you and shining over your shoulder, illuminate the deep recesses of the cell. The eggs are translucent white, and resemble a miniature grain of rice. I recommend that you invest in an inexpensive pair of reading glasses. The magnification can really help you spot the eggs (even if you don't normally need reading glasses). Once you discover your first egg, it'll be far easier to know what you're looking for during future inspections.

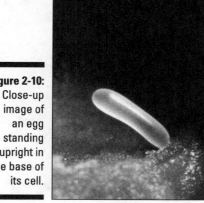

Figure 2-10:
Close-up
image of
an egg
standing
upright in
the base of
its cell.

Courtesy of Stephen McDaniel

Larva

Three days after the queen lays the egg, it hatches into a *larva* (the plural is *larvae*). Healthy larvae are snowy white and resemble small grubs curled up in the cells (see Figure 2-11). Tiny at first, the larvae grow quickly, shedding their skin five times. These helpless little creatures have voracious appetites, consuming 1,300 meals a day. The nurse bees first feed the larvae royal jelly and later they're weaned to a mixture of honey and pollen (sometimes referred to as *bee bread*). Within just five days, they are 1,570 times larger than their original size. At this time the worker bees seal the larvae in the cell with a porous capping of tan beeswax. Once sealed in, the larvae spin a cocoon around their bodies.

Pupa

The larva is now officially a *pupa* (the plural is *pupae*). Here's where things really begin to happen. Of course the transformations now taking place are hidden from sight under the wax cappings. But if you could, you'd see that this little creature is beginning to take on the familiar features of an adult bee (see Figure 2-12). The eyes, legs, and wings take shape. Coloration begins with the eyes. First pink, then purple, then black. Finally, the fine hairs that cover the bee's body develop. After 12 days, the now *adult bee* chews her way through the wax capping to join her sisters and brothers. Figure 2-13 shows the entire life cycle of the three castes of honey bee from start to finish.

Figure 2-11:
Beautiful
little larvae
curled up in
their cells.

Courtesy of John Clayton

Figure 2-12:
Opened
cells reveal
an egg and
developing
pupae.

Courtesy of Dr. Edward Ross, California Academy of Sciences

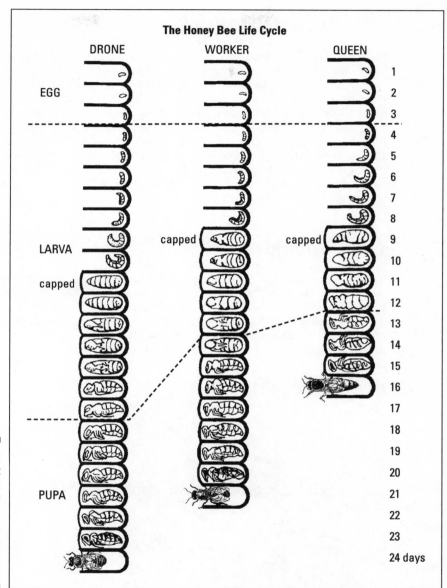

Figure 2-13:
This chart shows the daily development cycle of all three castes, from egg to adult.

Other Stinging Insects

Many people are quick to say they've been "stung by a bee," but the chances of a *honey bee* stinging them are rather slim. Honey bees usually are gentle in nature, and it is rare for an individual to be stung by a honey bee. Away from their hives, honey bees are nonaggressive. More aggressive insects are the more likely culprits when someone is stung. Most folks, however, don't make the distinction between honey bees and everything else. They incorrectly lump all insects with stingers into the "bee" category. True bees are unique in that their bodies are covered with hair, and they use pollen and nectar from plants as their sole source of food (they're not the ones raiding your cola drink at a picnic — those are likely to be yellow jackets). Here are some of the most common stinging insects.

Bumblebee

The gentle bumblebee (see Figure 2-14) is large, plump, and hairy. It's a familiar sight, buzzing loudly from flower to flower, collecting pollen and nectar. Bumblebees live in small ground nests that die off every autumn. At the peak of summer, the colony is only a few hundred strong. Bumblebees make honey, but only small amounts (measured in ounces, not pounds). They are docile and not inclined to sting, unless their nest is disturbed.

Figure 2-14: The bumblebee is furry and plump.

Courtesy of Dr. Edward Ross, California Academy of Sciences

Carpenter bee

The carpenter bee (see Figure 2-15) looks much like a bumblebee, but its habits are quite different. It is a solitary bee that makes its nest by tunneling through solid wood (sometimes the wooden eaves of a barn or shed). Like the honey bee, the carpenter bee forages for pollen. Its nest is small and produces only a few dozen offspring a season. Carpenter bees are gentle and are not likely to sting.

Figure 2-15: The carpenter bee looks similar to a bumblebee, but its abdomen has no hair.

Courtesy of Dr. Edward Ross, California Academy of Sciences

Wasp

Many different kinds of insects are called "wasps." The more familiar of these are distinguished by their smooth hard bodies (usually brown or black) and familiar ultra-thin "wasp waist" (see Figure 2-16). So-called "social wasps" build exposed paper or mud nests, which usually are rather small and contain only a handful of insects and brood. These nests sometimes are located where we'd rather not have them (in a door frame or windowsill). The slightest disturbance can lead to defensive behavior and stings. Social wasps primarily are meat eaters, but adult wasps are attracted to sweets.

Yellow jacket

The yellow jacket also is a social wasp. Fierce and highly aggressive, it is likely responsible for most of the stings wrongly attributed to bees (see Figure 2-17). Yellow jackets are a familiar sight at summer picnics where they scavenge for food and sugary drinks. Two basic kinds of yellow jackets exist: those that build their nests underground (which can create a problem when noisy lawn mowers or thundering feet pass overhead) and those that make their nests in trees. All in all, yellow jackets aren't very friendly bugs.

Figure 2-16: The wasp is clearly identified by its smooth hairless body and narrow "wasp waist."

Courtesy of Dr. Edward Ross, California Academy of Sciences

Figure 2-17: The ill-tempered yellow jacket is a meat eater but also has a taste for sweets.

Courtesy of Dr. Edward Ross, California Academy of Sciences

Bald-faced hornet

Bald-faced hornets are not loveable creatures (see Figure 2-18). They are related to yellow jackets, but they build their nests *above* ground. Hornets have a mean disposition and are ruthless hunters and meat eaters. They do, however, build fantastically impressive and beautiful paper nests from their saliva and wood fiber they harvest from dead trees (see Figure 2-19). These nests can grow large during the summer and eventually reach the size of a basketball. Such nests can contain several thousand hot-tempered hornets — keep your distance! In nontropical regions, the end of the summer marks the end of the hornet city. When the cool weather approaches, the nest is abandoned and only the queen survives. She finds a warm retreat underground and emerges in the spring, raising young and building a new nest.

Figure 2-18:
The bald-faced hornet makes impressive paper nests in trees.

Courtesy of Dr. Edward Ross, California Academy of Sciences

Figure 2-19:
A large paper nest made by a colony of bald-faced hornets.

Part II
Starting Your Adventure

The 5th Wave By Rich Tennant

"Honey - get the smoker. The guests are starting to swarm all over the bee hives again."

In this part . . .

This is where the fun begins! In these chapters, I tell you how to get started with honey bees, where you should locate your hive, and what kind of equipment you'll need. I also show you how to successfully and safely transfer your bees to their new home.

Chapter 3

Alleviating Apprehensions and Making Decisions

. . . .

In This Chapter

▶ Avoiding the dreaded stinger

▶ Understanding local restrictions

▶ Winning over your family, friends, and neighbors

▶ Deciding whether you have enough space

▶ Picking the perfect location

▶ Deciding the best time to start

. . . .

I suspect all new backyard beekeepers think similar thoughts as they're deciding to make the plunge. You've thought about beekeeping for some time. You're growing more and more intrigued by the idea . . . maybe this is the year you're going to do something. It certainly sounds like a lot of fun. What could be more unique? It's educational and a nice outdoor activity for you — back to nature and all that stuff. The bees will do a great job of pollinating the garden and there's that glorious crop of delicious homegrown honey to look forward to. The anticipation is building daily and you're consumed with excitement. That's it! You've made up your mind. You'll become a beekeeper! But in the back of your mind some nagging concerns keep bubbling to the surface.

You're a wee bit concerned about getting stung, aren't you? Your friends and family may say you're crazy for thinking of this. What if the neighbors disapprove when they find out? Maybe bees are not even *allowed* in your neighborhood. What happens if the bees don't like their new home and all fly away? Help!

These are certainly some of the concerns that I had when I first started. In this chapter I hope to defuse your apprehensions and suggest some helpful ways to deal with those concerns.

Overcoming Sting Phobia

Perhaps the best-known part of the bee's anatomy is its stinger. Quite honestly, that was my biggest apprehension about taking up beekeeping. I don't think I'd ever been stung by a honey bee, but I'd certainly felt the wrath of yellow jackets and hornets. I wanted no part of becoming a daily target for anything so unpleasant. I fretted about my fear for a long time, looking for reassurances from experienced beekeepers. They told me time and again that honey bees bred for beekeeping were docile and seldom inclined to sting. But lacking first-hand experience, I was doubtful.

The advice turned out to be 100 percent correct. Honey bees *are* docile and gentle creatures. To my surprise (and delight), I made it through my entire first season without receiving a single sting. In the nearly 20 years that I've been keeping bees, not a single member of my family, not a single visitor to my home, and not a single neighbor has ever been stung by one of my honey bees.

By the way, bees *sting* — they don't *bite*. Honey bees use their stinger only as a last resort to defend the colony. After all, they die after stinging. Away from the hive (while they're collecting nectar and pollen) defending the colony is no longer a priority, so they're as gentle as lambs.

Do I ever get stung? Sure. But usually not more than three or four times a year. In every case, the stings I take are a result of my own carelessness. I'm rushing, taking short cuts, or am inattentive to their mood — all things that I shouldn't do. The secret to avoiding stings is your technique and demeanor.

Here are some helpful tips for avoiding stings:

- Always wear a veil and use your smoker when visiting your hive (see Chapters 4 and 5 for more information on these two vital pieces of beekeeper apparatus).

- Inspect your bees during pleasant daytime weather. Try to use the hours between 10 a.m. and 5 p.m. That's when most of the bees are out working, and fewer bees are at home. Don't open up the hive at night, during bad weather, or if a thunderstorm is brewing. In Chapters 6 and 7, I go into detail about how to open the hive and inspect the colony.

- Don't rush. Take your time and move calmly. Sudden movements are a no-no.

- Get a good grip on frames. If you drop a frame of bees, you'll have a memorable story to tell.

✔ Never swat at bees. Become accustomed to them crawling on your hands and clothing. They're just exploring. Bees can be gently pushed aside if necessary.

✔ When woodenware is stuck together with propolis, don't snap it apart with a loud "crack." The bees go on full alert when they feel sudden vibrations.

✔ Never leave sugar syrup or honey in open containers near the hive. Doing so can excite bees into a frenzy, and you may find yourself in the middle of it. It can also set off *robbing* — an unwelcome situation in which bees from other colonies attack your bees, robbing them of their honey. In Chapter 9 you'll find instructions on how to avoid robbing, and what to do when it happens.

✔ Keep yourself and bee clothing laundered. Bees don't like bad body odor. If you like to eat garlic, avoid indulging right before visiting your bees. Chapter 6 has some handy hygiene hints.

✔ Wear light-colored clothing. Bees don't seem to like dark colors.

Knowing what to do if you're stung

Be prepared to answer the following question from everyone who learns you're a beekeeper: "Do you ever get stung?" You'll hear this one a hundred times. An occasional sting is a fact of life for a beekeeper. Following the rules of the road, however, keeps stings to a minimum, or perhaps you'll get none at all. Yet, if a bee stings you or your clothing, calmly remove the stinger and smoke the area to mask the chemical alarm scent left behind. (This alarm pheromone can stimulate other bees to sting.) To remove the stinger, use your fingernail to *scrape* it off your skin. Notice, in Figure 3-1, how the barbed stinger (and some of the bee's innards) remains with the victim as the bee moves away. This bee's a goner.

Don't try to *pinch* the stinger off. That only squeezes the venom sack left behind by the bee and injects more venom.

Apply a cold compress and take an antihistamine tablet (such as Benadryl). Antihistamine creams also are available. Using this technique alleviates the swelling, itching, and discomfort.

Some folks swear by the effectiveness of baking-soda-and-water poultices for bee stings; other folks advocate meat tenderizer and wet tobacco poultices, respectively. These are "grandma recipes" that were used before we had the antidote that the medical profession endorses — over-the-counter antihistamines.

Courtesy of Stephen McDaniel

Figure 3-1:
Ouch! This bee has made the ultimate sacrifice by stinging this beekeeper.

Watching for allergic reactions

All bee stings hurt a bit, but not for long. Experiencing redness, swelling, and itching is completely natural. These are normal (not allergic) reactions. For a small percentage of individuals, more severe allergic or even toxic reactions can occur, including severe swelling beyond the immediate area of the sting, and shortness of breath. In the worst cases, reaction to bee stings can result in loss of consciousness or even death. The most severe reactions occur in less than 1 percent of the population. To put that in perspective, more people are killed by lightning each year than die from bee stings.

As a precaution against a guest having a severe reaction, I keep an EpiPen (see Figure 3-2) on hand. These emergency sting kits are available from your doctor by prescription. The kit automatically injects a dosage of epinephrine (adrenaline). But be careful. Liability issues can arise when injecting another person, so check with your doctor beforehand.

Building up a tolerance

Now this may sound strange, but many beekeepers (myself included) look forward to getting a few stings early in the season. No, we're not masochistic.

The more stings we get, the less the swelling and itching. For many, occasional stings actually build up a kind of tolerance. It still smarts, but the side effects disappear.

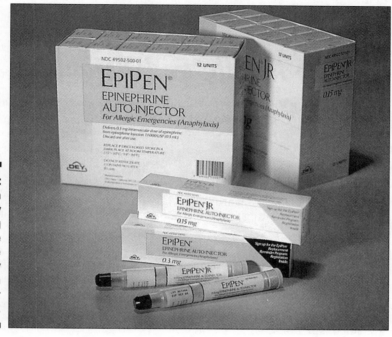

Figure 3-2:
EpiPen emergency sting kits are available only by prescription from your doctor.

One school of thought states that bee venom can actually be good for some health conditions that you may suffer from. This is what bee-sting therapy is all about; see Chapter 1 for more information.

Understanding Local Laws and Ordinances

Is it legal to keep bees? In most places, the answer is yes. But some areas have laws or ordinances restricting or even prohibiting beekeeping. For the most part, such restrictions are limited to highly populated, urban areas. Other communities may limit the number of hives you can keep, and some require you to register your bees with town hall. Some communities require that the state bee inspector inspect the health of your colonies periodically. If you have any questions about the legality of keeping bees, contact your state bee inspector, the state agricultural experiment station, or a local bee club or association. *Bee Culture* magazine maintains an online listing of "Who's Who in the Beeyard." This search engine is a great way to find beekeeping clubs, associations, and agencies in your state. Visit: www.beeculture.com.

Dealing with Family and Neighbors

For many among the general public, ignorance of honey bees is complete. Having been stung by hornets and yellow jackets, they assume having any kind of bee nearby spells trouble. So taking steps to educate them and alleviate their fears is up to you.

Some things you can do to put them at ease are

- Restricting your beeyard to two hives or less. Having a couple of hives is far less intimidating to the uneducated than if you had a whole phalanx of hives.

- Locating your hive in such a way that it doesn't point at your neighbor's driveway, your house entrance, or some other pedestrian traffic-way. Bees fly up, up, and away as they leave the hive. Once they're 15 feet from the hive, they're way above head level.

- Not flaunting your hives. Put them in an area where they'll be inconspicuous.

- Painting or staining your hives to blend into the environment.

- Providing a nearby source of water for your bees. That keeps them from collecting water from your neighbor's pool or birdbath (see the "Providing for your thirsty bees" section later in this chapter).

- Inviting folks to stop by and watch you inspect your hive. They'll see first-hand how gentle bees are, and your own enthusiasm will be contagious.

- Letting your neighbors know that bees fly in about a three-mile radius of home plate (that's roughly 6,000 square acres). So they'll be visiting a huge area that isn't anywhere near your neighbor's property.

- Giving gifts of honey to all your immediate neighbors (see Figure 3-3 for an example). This gesture goes a long ways in the public relations department.

Figure 3-3: This gift basket of honey bee products will be given to each of my immediate neighbors. That's sure to help keep the peace.

Deciding Where to Locate Your Hives

You can keep bees just about anywhere: in the countryside, in the city, in a corner of the garden, by the back door, in a field, on the terrace, or even on a rooftop downtown. You don't need a great deal of space, nor do you need to have flowers on your property. Bees will happily travel for miles to forage for what they need. These girls are amazingly adaptable, but you'll get optimum results and a more rewarding honey harvest if you follow some basic guidelines (see Figure 3-4). Basically, you're looking for easy access (so you can tend to your hives), good drainage (so the bees don't get wet), a nearby water source for the bees, dappled sunlight, and minimal wind. Keep in mind that fulfilling *all* these criteria may not be possible. Do the best you can by:

- Facing your hive to the southeast. That way your bees get an early morning wake-up call and start foraging early.

- Positioning your hive so that it is easily accessible come honey harvest time. You don't want to be hauling hundreds of pounds of honey up a hill on a hot August day.

- Providing a windbreak at the rear of the hive (see Figure 3-5). I've planted a few hemlocks behind my hives. Or you can erect a fence made from posts and burlap, blocking harsh winter winds that can stress the colony.

- Putting the hive in dappling sunlight. Ideally, avoid full sun, because the warmth of the sun requires the colony to work hard to regulate the hive's temperature in the summer. By contrast, you also want to avoid deep, dark shade, because it can make the hive damp and the colony listless.

- Making sure the hive has good ventilation. Avoid placing it in a gully where the air is still and damp. Also, avoid putting it at the peak of a hill, where bees are subject to winter's fury.

✔ Placing the hive absolutely level from side to side, and with the front of the hive just slightly lower than the rear (a difference of an inch or less is fine), so that any rainwater drains out of the hive (and not into it).

✔ Locating your hive on firm, dry land. Don't let it sink into the quagmire.

Mulch around the hive prevents grass and weeds from blocking its entrances.

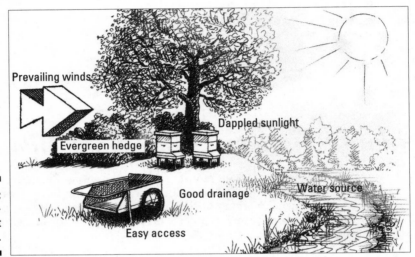

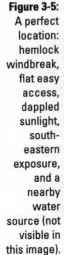

Figure 3-4:
The picture-perfect beeyard.

Figure 3-5:
A perfect location: hemlock windbreak, flat easy access, dappled sunlight, south-eastern exposure, and a nearby water source (not visible in this image).

How to move a full hive

It's best not to move hives around unless it's necessary because it's disruptive to the bees and a lot of work for you. But sometimes move you must. Here are some helpful guidelines:

✔ Plan to make your move in the evening when the bees are not flying.

✔ Before making the move, tape up any extra entrance or ventilation holes you have drilled in the hive (duct tape works great).

✔ Secure the hive together by using heavy-duty strapping tapes (available at hardware stores). These strapping tapes use a ratchet-type buckle to tighten the straps. Strap the entire hive together as a single unit: bottom board, hive bodies, and cover.

✔ Staple a strip of window screening across the front entrance of the hive. Doing so will keep the bees from flying out of the hive (and stinging you) while providing them with adequate ventilation.

✔ Use a hand truck to move the hive (an entire hive can weigh a couple hundred pounds). Get a friend to help.

✔ Wear a veil and gloves in case any bees get loose. I can assure you that they won't be happy about this move.

✔ Once the hive is in its new location, wait until early the next morning to remove the straps and the entrance screen. This gives the bees time to calm down.

Providing for your thirsty bees

During their foraging season, bees collect more than just nectar and pollen. They gather a whole lot of water. They use it to dilute honey that's too thick, and to cool the hive during hot weather. Field bees bring water back to the hive and deposit it in cells, while other bees fan their wings furiously to evaporate the water and regulate the temperature of the hive.

If your hive is at the edge of a stream or pond, that's perfect. But if it isn't, you should provide a nearby water source for the bees. Keep in mind that they'll seek out the nearest water source. You certainly don't want that to be your neighbor's kiddy pool. You can improvise all kinds of watering devices. Figure 3-6 shows an attractive and natural-looking watering device that I created on top of a boulder that sits in one of my beeyards. All it took was a little cement, a dozen rocks and a few minutes of amateur masonry skills.

Consider these other watering options: a pie pan filled with gravel and topped off with water, a chicken-watering device (available at farm supply stores; see Figure 3-7), or simply an outdoor faucet that is encouraged to develop a slow drip.

Figure 3-6:
A shallow bee watering pool that I constructed on a boulder near my hives.

Figure 3-7:
A chicken waterer is a great way to provide your bees with water. Place some gravel or small pebbles in the tray, so the bees don't drown.

Understanding the correlation between geographical area and honey flavors

The type of honey you eat usually is classified by the primary floral sources from which the bees gathered the nectar. A colony hived in the midst of a huge orange grove collects nectar from the orange blossoms — thus the bees make orange-blossom honey. Bees in a field of clover make clover honey, and so on. As many different kinds of honey can exist as there are flowers that bloom. The list gets long.

Honey bee wet bar

When it comes to providing water for your bees, here's a nifty idea that I learned from a fellow beekeeper. Find or purchase a clean pail or bucket. Any size, color or material will do. Just make sure that it's clean and has never been used for chemicals, fertilizers, or pesticides. Drill ½-inch drainage holes all around the top edge of the bucket. The holes should be placed about 2 to 3 inches down from the top. Fill the bucket nearly to the holes with water, and then float a single layer of Styrofoam packaging pellets on the surface of the water. The pellets give the bees something to stand on as they sip water. That way they won't drown. The drainage holes keep rainwater from overflowing the bucket and washing away the pellets. Neat, huh?

For most hobbyists, the flavor of honey they harvest depends upon the dominant floral sources in their areas. During the course of a season, your bees visit many different floral sources. They bring in many different kinds of nectar. The resulting honey, therefore, can properly be classified as *wildflower honey,* a natural blend of various floral sources.

The beekeeper who is determined to harvest a particular kind of honey (clover, blueberry, apple blossom, sage, tupelo, buckwheat, and so on) needs to locate his or her colony in the midst of acres of this preferred source and must harvest the honey as soon as that desired bloom is over. But, doing so is not very practical for the backyard beekeeper. That's what the commercial beekeepers do.

My advice? Let the bees do their thing and collect from myriad nectar sources. You'll not be disappointed in the resulting harvest, because it will be unique to your neighborhood and better than anything you have ever tasted from the supermarket.

The color of honey

Honey can be all different colors. Hundreds of unique kinds and colors of honey range from water-white to black-amber (and every hue in between). Even a "blue" honey is derived from sourwood blossoms in eastern North Carolina. The source of the nectar determines the color and flavor of honey. Usually, the darker the color, the stronger and more intricate the flavor and aroma.

Knowing the Best Time to Start Your Adventure

In a sense, winter is the best time to start. That's a good time to order and assemble the equipment that you'll need and to reserve a package of bees for early spring delivery. Use the winter months to read up on bees and beekeeping and become familiar with your equipment. Join a bee club and attend their meetings. That's a great way to learn more about beekeeping and meet new friends. Many clubs have special programs for new beekeepers and hands-on weekend workshops that show you how it's done. Latch onto a mentor whom you can call on to answer questions and help you get started.

Install your bees in the early spring (April or early May is best). Spring varies from area to area, but you're trying to time your start to coincide with the first early season blossoms, and just a few weeks prior to the fruit bloom. Don't wait until June or July. Starting a hive in summer won't give your colony a chance to grow strong for its first winter.

Be sure to have everything assembled and ready to go before the post office calls announcing the arrival of your bees. As for what kind of equipment you need to get for this new adventure, that's covered in Chapter 4.

Chapter 4

Basic Equipment for Beekeepers

● ●

In This Chapter
▶ Deciding what equipment and tools to get

▶ Assembling woodenware

▶ Preserving the wood to last for years

▶ Learning some tricks of the trade

▶ Really helpful accessories

● ●

*A*ll kinds of fantastic tools, gadgets and equipment are used by beekeepers. Quite frankly, part of the fun of beekeeping is putting your hive together and using the paraphernalia that goes with it. The makings for a beehive come in a kit form and are precut to make assembly a breeze. The work is neither difficult, nor does it require too much skill.

The more adventuresome among you may want to try making your own hives from scratch. But precise measurements are critical to the bees, and unless you're pretty good at carpentry and have a lot of time to spend, purchasing what you need is probably easier. Once you get the hang of beekeeping, you can try your hand at making your own hives.

Many different mail order establishments offer beekeeping supplies, and a number of excellent ones are now on the Internet. Check out a listing of some of the quality suppliers in the Appendix A.

Finding Out about the Langstroth Hive

Many different kinds and sizes of beehives are available. But worldwide, the most common is the 10-frame *Langstroth hive*. This so-called moveable frame hive with a practical top opening was the 1851 invention of Rev. Lorenzo L. Langstroth of Pennsylvania (see Figure 4-1). Its design hasn't changed much in the last 150 years, which is a testament to its practicality.

Figure 4-1:
The "father" of modern beekeeping, Lorenzo Langstroth.

Here are some of the benefits of the Langstroth hive:

✔ Langstroth hive parts are completely interchangeable and readily available from any beekeeping supply vendor.

✔ All interior parts of the hive are spaced exactly three-eighths of an inch apart (9.525 mm), thus enabling honey bees to build straight and even combs. Because it provides the right "bee space," the bees don't "glue" parts together with propolis or burr comb.

✔ Langstroth's design enables beekeepers to freely inspect and manipulate frames of comb. Prior to this discovery, beekeepers were unable to inspect hives for disease, and the only way to harvest wax and honey was to kill the bees or drive them from the hive.

A popular early hive

This simple basket hive, or *skep* (depicted in this book's Hive Talk icon) was popular for hundreds of years in many countries. But, with this design, you have no way to inspect the bees' health and no way to harvest honey without destroying the bees and comb. Although the skep hive is rarely used today, it still is associated with the public's "romantic" image of what a beehive looks like.

Knowing the Basic Woodenware Parts of the Hive

Woodenware refers to the various components that collectively result in the beehive. Traditionally these components are made of wood — thus the term — but some manufacturers offer synthetic versions of these same components (plastic, polystyrene, and so on). My advice: Get the wood. The bees accept it far more readily than synthetic versions. And the smell and feel of wood is ever so much more pleasurable to work with.

Be aware that the hive parts you order (see "Ordering Hive Parts" later in this chapter) will arrive in precut pieces. You will need to spend some time assembling them. See "Setting Up Shop" later in this chapter for a list of tools and so forth that you will need for assembly. **Note:** For an additional charge, some vendors will preassemble hives for you.

This section discusses, bottom to top, the various components of a modern Langstroth beehive. As you read this section, refer to Figure 4-2 to see what the various parts look like and where they are located within the structure of the hive.

Hive stand

The entire hive sits on a hive stand. The best ones are made of cypress — a wood that is highly resistant to rot. The stand is an important component of the hive because it elevates the hive off the ground, improving circulation and minimizing dampness. In addition, grass growing in front of the hive's entrance can slow the bees' ability to get in and out. The stand alleviates that problem by raising the hive above the grass.

The hive stand consists of three rails and a landing board, upon which the bees land when they return home from foraging trips. Nailing on the landing board just right is the only tricky part of hive stand assembly. Carefully follow the instructions that come with your hive stand. **Note:** Putting the stand together on a flat surface helps prevent the stand from wobbling.

Bottom board

The *bottom board* is the thick bottom floor of the beehive. Like the hive stand, the best bottom boards are made of cypress wood. This part's easy and intuitive to put together.

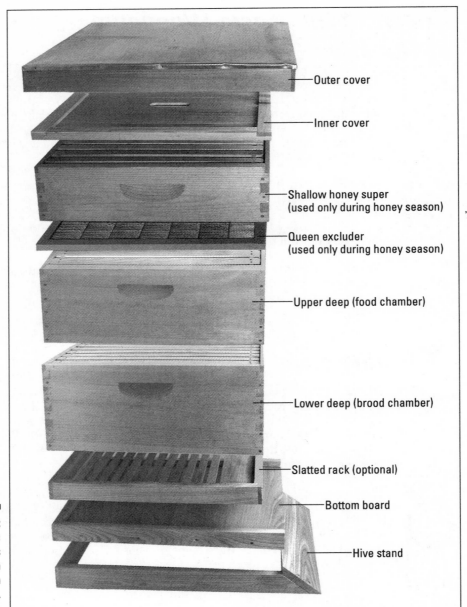

Figure 4-2:
The basic
components
of a modern
Langstroth
hive.

Outer cover

Inner cover

Shallow honey super
(used only during honey season)

Queen excluder
(used only during honey season)

Upper deep (food chamber)

Lower deep (brood chamber)

Slatted rack (optional)

Bottom board

Hive stand

Entrance reducer

When you order a bottom board it comes with a notched wooden cleat. The cleat serves as your *entrance reducer,* which limits bee access to the hive and controls ventilation and temperature during cooler months. The entrance reducer isn't nailed into place, but rather is placed loosely at the hive's entrance. The small notch reduces the entrance of the hive to the width of a finger. The larger notch opens the entrance to about four finger widths. Removing the entrance reducer completely opens the entrance.

Beekeepers use the entrance reducer only for newly established hives or during cold weather (see Chapter 5). This is the reason the entrance reducer isn't shown in Figure 4-2. For established hives in warm weather, the entrance reducer isn't used at all. The only exception may be when you're dealing with a robbing situation — see Chapter 9.

Deep-hive body

The *deep-hive body* contains ten frames of honeycomb. The best quality ones are made of clear pine or cypress and have crisply cut dovetail joints for added strength. You'll need two deep-hive bodies to stack one on top of the other, like a two-story condo. The bees use the *lower deep* as the nursery, or *brood chamber,* to raise thousands of baby bees. The bees use the *upper deep* as the pantry or *food chamber,* where they store most of the honey and pollen for their use.

The hive body assembles easily. It consists of four precut planks of wood that come together to form a simple box. Simply match up the four planks, and drive nails into the predrilled holes. Hammering a single nail in the center of each of the four joints keeps the box square. Use a carpenter's square to even things up before hammering in the remaining nails.

Place the hive body on the bottom board. If it rocks or wobbles a little, use some coarse sandpaper or a plane to remove any high spots. The hive body needs to fit solidly on the stand.

Use a little waterproof wood glue on the joints of all your woodenware before nailing them together. That gives you a super-strong bond.

Queen excluder

No matter what style of honey harvest you choose, a queen excluder is a basic piece of equipment you need. It's placed between the deep food chamber and the shallow honey supers. The queen excluder comes already assembled and consists of a wooden frame holding a grid of metal wire, or a perforated sheet of plastic (see Figure 4-3). As the name implies, this gizmo prevents the queen from entering the honey super and laying eggs. Otherwise, a queen laying eggs in the super encourages bees to bring pollen into the super, spoiling the clarity of the honey. The spacing of the grid is such that smaller worker bees can pass through to the honey supers.

You use a queen excluder only when you are collecting honey. It is a piece of woodenware that is unique to honey production. When you are not collecting honey it should not be used.

Figure 4-3:
A queen
excluder.

For the do-it-yourselfer

If you're handy, you may want to try building your own equipment. For the more adventuresome, I include some plans in Chapter 14 to help you along. Remember that precise measurements are critical within a hive. Bees require a precise *bee space*. If you wind up with too little space for the bees, they'll glue everything together with propolis. Too much space, and they'll fill it with burr comb. Either way, it makes the manipulation and inspection of frames impossible. So, measure carefully.

Making your woodenware last

To get the maximum life from your equipment, you must protect it from the elements. If you don't, wood that's exposed to the weather rots.

After you've assembled your equipment and before you put your bees in their new home, paint all the *outer* surfaces of the hive (see the following list). Use (at least) two coats of a good quality outdoor paint (either latex or oil-based paints are okay). The color is up to you. Any light pastel color is fine, but avoid dark colors, because they will make the hive too hot during summer. White seems to be the most traditional color. If you prefer, you can stain your woodenware and treat it with an outdoor grade of polyurethane.

Do paint/treat the following:

✔ Hive-top feeder (outside surfaces only) ?

✔ Outer cover (inside and outside surfaces)

✔ Supers and hive bodies (outside surfaces only)

✔ Bottom board (all surfaces)

✔ Hive stand (all surfaces)

Do not paint/treat the following:

✔ Inner cover

✔ Frames

✔ Inside surfaces of supers, hive bodies, and hive-top feeder

✔ Queen excluder

Shallow honey super

Shallow honey supers are used by beekeepers to collect surplus honey. That's *your* honey — the honey that you can harvest from your bees. The honey that's in the deep-hive body you need to leave for the bees. Supers are identical in design to the deep-hive bodies — and assemble in a similar manner — but the depth of the supers is shallower.

Honey supers are put on the hive about eight weeks after you first install your bees. For the second-year beekeeper, honey supers are placed on the hive when the first spring flowers start to bloom.

The shallow depth of the supers makes them easy to handle during the honey harvest. A shallow super full of honey will weigh a hefty (but manageable) 40 pounds. However a *deep-hive body* full of honey weighs a backbreaking 80 pounds. That's more weight than you'd want to deal with!

 As the bees collect more honey, you can add more honey supers to the hive, stacking them one on top of another like so many stories to a skyscraper. See "Ordering Hive Parts" later in this chapter for information on how many supers you should have on hand for your first year.

Frames

Each wooden frame contains a single sheet of beeswax foundation (described in the next section). The frame is kind of like a picture frame. It firmly holds the wax and enables you to remove these panels of honeycomb for inspection or honey extraction. Ten deep frames are used in each deep-hive body, and nine shallow frames usually are used in each shallow honey super. Frames are the trickiest pieces of equipment you'll have to assemble. Beekeeping suppliers usually sell frames in packages of ten, with hardware included.

Frames come in two basic sizes: deep and shallow — corresponding to deep-hive bodies and shallow honey supers. The method for assembling deep and shallow frames is identical. Regardless of its size, each frame has four basic components: one top bar with a wedge (the wedge holds the foundation in place), one grooved-bottom bar, and two sidebars (see Figure 4-4). Frames typically are supplied with the necessary and correct size nails.

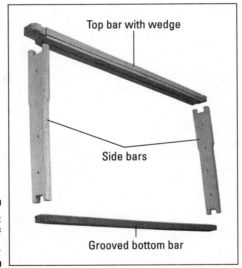

Top bar with wedge

Side bars

Figure 4-4:
The parts of
a frame.

Grooved bottom bar

Nine or ten frames?

I like using nine frames in my honey supers. I also use special spacers along the frame rails to keep the distance between frames exact. Why do I use nine? That little extra space between each frame allows the bees to draw the comb much deeper. This results in more honey in the nine frames than there would have been in ten.

Assemble your frames by following these directions:

1. Take the top bar and snap out the wedge strip. You can use your hive tool to pry the wedge strip from its place. Clean up any *filigree* (rough edges) by scraping the wood with your hive tool. Save the wedge for use when you're installing the wax foundation (see "Foundation" section next).

2. Place the top bar on your tabletop work surface with the flat side facing down on the table.

3. Take the two side pieces and snap the wider end into the slots at either end of the top bar.

4. Now snap your bottom bar into the slots at the narrow end of the side pieces.

5. Now nail all four pieces together. Use a total of six nails per frame (two for each end of the top bar, and one at each end of the bottom bar). In addition to nailing, I suggest that you also glue the parts together using an all-weather wood glue. Doing so adds strength.

6. Repeat these steps until all your frames are assembled. Time for a break while the glue dries.

Don't be tempted to use any shortcuts. Frames undergo all kinds of abuse and stress, so their structural integrity is vital. Use glue for extra strength and don't skimp on the nails nor settle for a bent nail that's partially driven home. There's no cheating when it comes to assembling frames!

Foundation

Foundation consists of thin rectangular sheets that are used to urge your bees to draw even and uniform honeycombs. It comes in two forms: plastic and beeswax. Using plastic foundation has some advantages, because it's stronger than wax and resists wax moth infestations. But the bees are slow to accept plastic, and I don't recommend it for the new beekeeper. Instead, purchase foundation made from pure beeswax.

Beeswax foundation is vertically wired for strength, and imprinted with a hexagonal cell pattern that guides the bees as they draw out uniform, even combs. Your bees find the sweet smell of beeswax foundation irresistible and quickly draw out each sheet into 7,000 beautiful, uniform cells (3,500 cells on each side) where they can store their food, raise brood, and collect honey for you!

Like frames, foundation comes in deep and shallow sizes — deep for the deep-hive bodies and shallow for the shallow supers. You insert the foundation into the frames the same way for both.

Here's how to insert foundation into your frames:

1. With one hand, hold the frame upright on the table. Look closely at a sheet of foundation. Note that vertical wires protrude from one side and are bent at right angles. However, wires at the other side are trimmed flush with the foundation. Drop this flush end into the long groove of the bottom bar and then coax the other end of the foundation into the space where the wedge bar was (see Figure 4-5).

2. Turn the frame and foundation upside down (with the top bar now resting flat on the table). Adjust the foundation laterally so that equal space is on the left and right. Remember the wedge strip you removed when assembling the frames . . . now's the time to use it! Return the wedge strip to its place, sandwiching the foundation's bent wires between the wedge strip and the top bar (see Figure 4-6). Use a brad driver to nail the wedge strip to the top bar (see Figure 4-7). Start with one brad in the center, and then one brad at each end of the wedge strip. Add two more brads for good luck (five total).

Finally, use support pins — they look like little metal clothespins — to hold the foundation securely in place (see Figure 4-8). The pins go through predrilled holes in the side bars, and pinch the foundation to hold it in place. Although each side bar has three to four predrilled holes, use only *two* pins on each sidebar (4 per frame).

That's it! You've completed building one frame. Only 19 more to go!

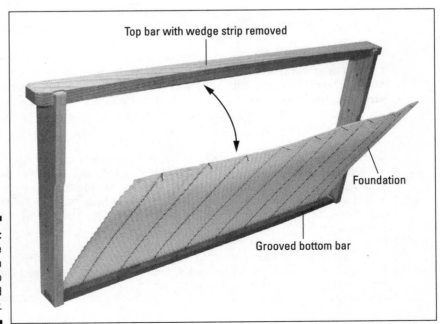

Figure 4-5:
Inserting the foundation sheet into the grooved bottom bar.

Top bar with wedge strip removed

Foundation

Grooved bottom bar

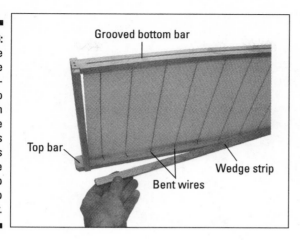

Figure 4-6: Turn the frame upside-down to sandwich the foundation's bent wires between the wedge strip and the top bar.

Grooved bottom bar

Top bar

Bent wires

Wedge strip

Figure 4-7: With the frame still upside-down, use a brad driver to nail the wedge strip back in place.

Knowing the right and wrong way to put the inner cover on the hive

There is a correct and incorrect way to put the inner cover on the hive. Note that there's a completely flat side, and a side with a ridge on all four sides. One of these ridges has a ventilation notch cut out of it. The inner cover goes immediately under the outer cover. The side with the ridge faces skyward. The notched ventilation hole goes towards the front of the hive.

Note: You do *not* use the inner cover at the same time you have a hive-top feeder on the hive. The hive-top feeder (described later in this chapter) takes the place of the inner cover.

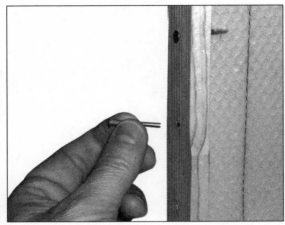

Figure 4-8:
Insert
support pins
in predrilled
holes to
hold the
wax
foundation
in place.

Inner cover

The best quality inner covers are made entirely of cypress wood. Budget models made from pressboard or Masonite also are available, but they don't seem to last as long. The basic design consists of a wooden frame that houses flat planks of wood. In the center is a precut oval hole. When assembled, the inner cover resembles a shallow tray (with a hole in the center). A notch is cut out of one of the lengths of frame. This ventilation notch is positioned to the front of the hive. The inner cover is placed on the hive with the "tray" side facing up.

Outer cover

Look for cypress wood when buying an outer cover. Cypress resists rot and lasts the longest. Outer covers assemble in a manner similar to the inner cover: a frame containing flat planks of wood. But the *outer* cover has a galvanized steel tray that fits on the top, protecting it from the elements.

Ordering Hive Parts

Hive manufacturers traditionally make their woodenware out of pine and/or cypress. Hardwoods are fine, but too expensive for most hobbyists. A custom mahogany hive, for instance, runs over $1000 versus a standard pine and cypress hive for about $100. Many suppliers offer various grades of components from a *commercial* budget-grade to a *select* best quality-grade. Go for the highest quality that your budget allows. Although they may be a little more expensive upfront, quality parts assemble with greater ease, and are far more likely to outlast the budget versions.

Any of this stuff is available from beekeeping supply stores. Many of these vendors are now on the web. For a listing of some of my favorites, see Appendix A.

Startup hive kits

Many suppliers offer a basic startup kit that takes the guesswork out of what you need to get. These kits often are priced to save a few bucks. Make certain that your kit contains these basic items, discussed in this section:

✔ Bottom board

✔ Lower and upper deep

✔ Honey super

✔ Inner and outer covers

✔ Frames and foundation for both deeps and the honey super

✔ Hardware to assemble stuff (various size nails, foundations pins, and so on)

✔ Veil and gloves

✔ Smoker

✔ Hive tool

Anticipating the length of assembly time you'll need

By all means make sure every thing is ready *before* your bees arrive on your doorstep. Don't wait until the last minute to put things together. It probably will take a bit longer than you think, particularly if you are doing this for the first time.

First-timers should allow 2-3 hours to assemble the two deep-hive bodies, bottom board, and the outer and inner cover. Assembling frames and installing foundation may require another few hours. And then you need to allow an hour or two to paint your equipment. Plus there's the cleanup time and the time for the paint to dry. All in all, your weekend is cut out for you.

My advice? Get your equipment several weeks before your bees are scheduled to arrive. Use those few weeks to leisurely put things together.

Setting Up Shop

Before your bees arrive, you'll need to order and assemble the components that will become their new home. You don't need much space for putting the equipment together. A corner of the garage, basement, or even the kitchen will do just fine. A worktable is mighty handy, unless you actually like crawling around on the floor.

Get all your hive parts (woodenware) together and the instruction sheets that come with them. The only tool you absolutely *must* have is a hammer. But having the following also is mighty useful:

- ✔ A pair of pliers to remove nails that bend when you try to hammer them.
- ✔ A brad driver with some ¾-inch, 18-gauge brads. Having this tool makes the installation of wax foundation go much faster.
- ✔ A bottle of good quality all-weather wood glue. Gluing and nailing woodenware greatly improves their strength and longevity.
- ✔ A carpenter's square to ensure parts won't wobble when assembled.
- ✔ Some coarse sandpaper or a plane to tidy-up any uneven spots.
- ✔ A hive tool (I hope one came with your start-up kit). It's pretty handy for pulling nails and prying off the frame's wedge strip.

Start assembling your equipment from the ground up. That means starting with the hive stand, moving on to the bottom board, and so forth.

The various assembled parts of the hive are not nailed together. They simply are stacked one on top of the other (like a stack of pancakes). This enables you to open up and manipulate the hive and its parts during inspections.

Adding on Feeders

Feeders are used to offer sugar syrup to your bees when the nectar flow is minimal or nonexistent. They also provide a convenient way to medicate your bees (some medications can be dissolved in sugar syrup and fed to your bees). You must feed and medicate your bees in the early spring and once again in autumn (see Chapter 8). Each of the many different kinds of feeders has its pluses and minuses. I've included a brief description of the more popular varieties.

Hive-top feeder

The hive-top feeder is the model I urge you to use (see Figure 4-9). As a new beekeeper, you will love how easy and safe it is to use. The hive-top feeder sits directly on top of the upper deep brood box, and under the outer cover (no inner cover is used when a hive-top feeder is in place). It has a reservoir that can hold a gallon or two of syrup. Bees enter the feeder from below by means of a screened access.

The hive-top feeder has several distinct advantages over other types of feeders:

- ✔ Its large capacity means that you don't have to fill the feeder more than once every week or two.
- ✔ The screened bee access means that you can fill the feeder without risk of being stung (the bees are on the opposite side of the screen).
- ✔ Because you don't have to completely open the hive to refill it, you don't disturb the colony (every time you smoke and open a hive you set the bees' progress back a few days).
- ✔ Because the syrup is not exposed to the sun, you can add medication without concern that light will diminish its effectiveness.

Figure 4-9:
A hive-top
feeder.

Entrance feeder

The entrance feeder is a popular device consisting of a small inverted jar of syrup that sits in a contraption at the entrance to the hive (see Figure 4-10). Entrance feeders are inexpensive and simple to use. I don't recommend that you use an entrance feeder, however. They have some worrisome disadvantages:

- ✔ The feeder's proximity to the entrance can encourage bees from other hives to rob syrup and honey from your hive.
- ✔ You're unable to medicate the syrup because it sits directly in the sun.
- ✔ The feeder's exposure to the hot sun tends to spoil the syrup.
- ✔ Refilling the small jar frequently is necessary (often daily).
- ✔ Using an entrance feeder in the spring isn't effective. The entrance feeder is at the bottom of the hive, but the spring cluster of bees is at the *top* of the hive.
- ✔ Being at the entrance, you risk being stung by guard bees when you refill the feeder.

Figure 4-10: An entrance feeder.

Pail feeder

The pail feeder consists of a one-gallon plastic pail with a friction top closure. Several tiny holes are drilled in its top. The pail is filled with syrup and the friction top is snapped into place. The pail then is inverted and placed over the oval hole in the inner cover (Figure 4-11). A vacuum is created, and the syrup remains in the pail, yet is available to the bees that feed via the small holes. Although inexpensive and relatively easy to use, it also has a few disadvantages:

- ✔ This feeder is placed within an empty deep-hive body, with the outer cover on top.
- ✔ You essentially must open the hive to refill the feeder, leaving you vulnerable to stings.
- ✔ Refilling this feeder requires smoking your bees and disrupting the colony.
- ✔ Its one-gallon capacity requires refilling once or twice a week.
- ✔ Limited access to syrup means that only a few bees at a time can feed.

Figure 4-11:
Here's a pail feeder placed over the oval hole of the inner cover. By covering the feeder with an empty deep-hive body you can keep raccoons out of the feeder.

Frame feeder

This plastic feeder is a narrow vessel resembling a standard frame that is placed in the upper deep-hive body, replacing one of the wall frames (see Figure 4-12). Filled with a pint or two of syrup, bees have direct access to it. But it isn't practical:

✔ Its capacity is small and must be refilled frequently, two or three times a week.

✔ You lose the use of one frame while the feeder is in place.

✔ Opening the hive to refill the feeder is disruptive to the colony and exposes you to stings.

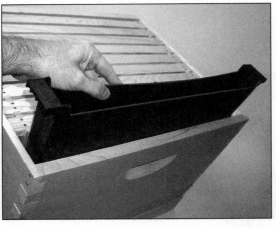

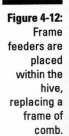

Figure 4-12: Frame feeders are placed within the hive, replacing a frame of comb.

Fundamental Tools

Two tools — the smoker and the hive tool — are a must for the beekeeper. They're used every time you visit the hive and are indispensable.

Smoker

The smoker will become your best friend. Smoke calms the bees and enables you to safely inspect your hive. Quite simply, the smoker is a fire chamber with bellows designed to produce lots of cool smoke. Smokers come in all shapes, sizes, and price ranges. The style that you choose doesn't really matter. Learn how to light it so that it stays lit, and never overdo the smoking process. A little smoke goes a long way. (See Chapter 6 for more about how to use your smoker.)

Hive tool

The versatility of the simple hive tool is impressive. Don't visit your hives without it! Use it to scrape wax and propolis off woodenware. Use it to loosen hive parts, open the hive, and manipulate frames. You can choose from various models (see Figure 4-13). To see pictures of the hive tool in action, go to Chapters 6 and 7.

Figure 4-13: Two varieties of hive tools.

Bee-Proof Clothing

New beekeepers should wear a long-sleeved shirt when visiting the hive. Light colors are best — bees don't like dark colors. Wear long pants and slip-on boots. Tuck your pant legs into the boots. Alternatively, use Velcro or elastic strips (even rubber bands) to secure your pant legs closed. You don't want a curious bee exploring up your leg! You should also invest in veils and gloves, which are discussed in this section.

Veils

Don't ever visit your hive without wearing a veil. Although your bees are likely to be gentle (especially during your first season), it defies common sense to put yourself at risk. Veils come in many different models (see Figure 4-14) and price ranges. Some are simple veils that slip over your head; others are integral to a pullover blouse or even a full jumpsuit. Pick the style that appeals most to you. If your colony tends to be more aggressive, more protection is advised. But remember, the more that you wear, the hotter you'll be during summer inspections. (See Chapter 6 for additional information on what to wear.)

Keep an extra veil or two on hand for visitors who want to watch while you inspect your bees.

Figure 4-14:
Protective clothing comes in various styles, from minimal to full coverage. This beekeeper uses a veil-and-blouse combination, leather gloves, and high boots to keep him *bee-tight.*

Gloves

New beekeepers like the idea of using gloves (see Figure 4-15), but I urge you not to use them for installing your bees or for routine inspections. You don't really need them at those times, especially with a new colony or early in the season. Gloves only make you clumsier. They inhibit your sense of touch, which can result in your inadvertently injuring bees. That's counterproductive and only makes them more defensive when they see you coming.

The only times that you need to use gloves are

- ✔ Late in the season (when your colony is at its strongest)
- ✔ During honey harvest season (when your bees are protective of their honey)
- ✔ When moving hive bodies (when you have a great deal of heavy work to do in a short period of time)

Other times leave the gloves at home. If you must, you can use heavy gardening gloves, or special beekeeping gloves with long sleeves (available from beekeeping supply vendors).

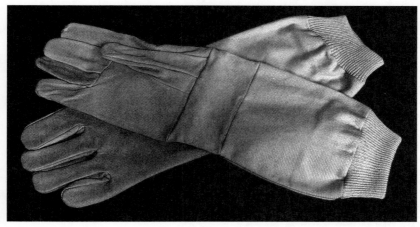

Figure 4-15: Although not needed for routine inspections, it's a good idea to have a pair of protective gloves.

Really Helpful Accessories

All kinds of gadgets, gizmos, and doodads are available to the beekeeper. Some are more useful than others. I describe a few of my favorites in this section.

Elevated hive stand

I have all my hives on elevated stands. Elevated hive stands are something you're more likely to build than purchase. The simplest elevated stands are made from four 14-inch lengths of two-by-four (use these for the legs) and a single plank of plywood that is large enough to hold the hive (see Figure 4-16). Put the entire hive on top of the elevated stand, raising it a little more than 14 inches off the ground. Alternatively, fashion an elevated stand from a few cinderblocks (see Figure 4-17). You can also use posts of various sorts (see Figure 4-18).

In either case, having the hive off the ground means no bending over during inspections. Doing so makes the hive far easier to work with. Elevating the entrance also helps deter skunks from snacking on your bees (see Chapter 11 for more on skunks and other kinds of pests).

Figure 4-16: You can build a simple table stand to elevate your hive off the damp ground.

Figure 4-17: You can also elevate your hive on cinder-blocks.

Figure 4-18:
You can use
a level
stump to get
your hive up
off the
ground.

Frame rest

A frame rest is a super-helpful device that I love. This product hangs on the side of the hive, providing a convenient and secure place to rest frames during routine inspections (see Figure 4-19). It holds up to three frames, giving you plenty of room in the hive to manipulate other frames without crushing bees.

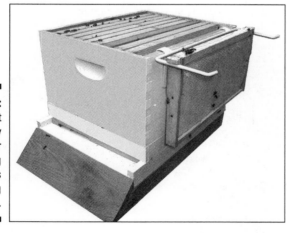

Figure 4-19:
A frame rest
is a handy
device for
holding
frames
during
inspections.

Bee brush

The long, super-soft bristles of a bee brush enable you to remove bees from frames and clothing without hurting them (see Figure 4-20). Some beekeepers use a goose feather for this purpose. Keep that in mind in the event you have an extra goose around the house.

Figure 4-20:
Use a soft bee brush to gently remove bees from frames and clothing.

Slatted rack

You might want to sandwich a slatted rack between the hive's bottom board and lower deep-hive body (see Figure 4-21). It does an excellent job of helping air circulation throughout the hive. Also, no cold drafts reach the front of the hive, which, in turn, encourages the queen to lay eggs right to the front of the combs. More eggs mean more bees, stronger hives, and more honey for you! I use a slatted rack on all my hives.

Figure 4-21:
Slatted racks help improve ventilation and can improve brood pattern.

Screened bottom board

With varroa mites becoming increasingly problematic for beekeepers, screened bottom boards are gaining popularity. (See Chapter 11 for more information on varroa mites.) A large percentage of mites naturally fall off the bees each day and land on the bottom board of the hive. Ordinarily they just crawl back up and reattach themselves to the bees, but not when you use a screened bottom board in place of a regular bottom board (see Figure 4-22). Mites drop off the bees and are trapped *under* the screening. They're unable to crawl back up into the hive. Studies have shown that using a screened bottom board can reduce the mite population by 30 percent without the use of any hazardous chemicals.

Figure 4-22: A screened bottom board can help control mites.

Other necessities

Some other necessities that all beekeepers should have on hand include:

- **A spray bottle of alcohol.** Fill a small plastic spray bottle with plain rubbing alcohol. Use this during inspections to clean any sticky honey or pollen off your hands. Never spray the bees with this!

- **Baby powder.** Dust your hands with baby powder before inspections. The bees seem to like the smell, and it helps keep your hands clean.

- **Toolbox.** Use a container to hold all your beekeeping tools. That way everything you need will be available to you during inspections. Any box will do. I use a fishing tackle box (see Figure 4-23).

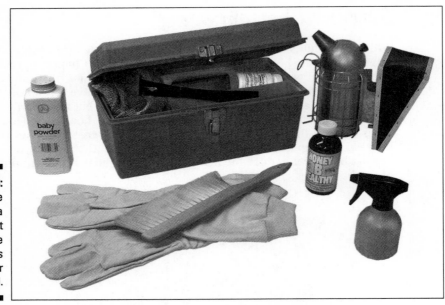

Figure 4-23:
A simple
toolbox is a
convenient
way to tote
supplies
to your
beeyard.

Chapter 5

Obtaining and Installing Your Bees

In This Chapter

▶ Knowing the kind of honey bee you want to raise

▶ Deciding how and where to obtain your bees

▶ Preparing for your bees' arrival

▶ Getting the girls into their new home

*O*rdering your bees and putting them into their new home (hiving) is just about my favorite part of beekeeping. *Hiving* your bees is surprisingly easy — and a lot safer than you might imagine. You don't often get an opportunity to do it, because once your bees are established you don't need to purchase a new colony. Bees are perennial and remain in their hive generation after generation. Only when you start a new hive or lose a hive to disease or starvation do you need to buy and install a new colony of bees.

I was a nervous wreck in the days and hours prior to installing my first colony. Like an expectant father, I paced the floor nervously until the day they arrived. And when they arrived, I fretted about how in the world I'd get all those bees into the hive. Would they fly away? Would they attack and sting me? Would the queen be okay? Would I do the right things? Help! All my fears and apprehensions turned out to be unfounded. It was as easy as pie, and a thoroughly delightful experience.

Determining the Kind of Bee You Want

You can choose from many different races and hybrids of honey bees. Each strain has its own pluses and minuses. The list below acquaints you with some of the more common types of bees. Most of these types are readily available from bee suppliers. Some suppliers even specialize in particular breeds, so shop around to find what you want.

- **Italian *(A. m. ligustica)*.** These honey bees are yellow-brown in color with distinct dark bands. This race originally hails from the Appenine Peninsula in Italy. They are good comb producers and the large brood that Italian bees produce results in quick colony growth. They maintain a big winter colony, however, which requires large stores of food. You can help offset this by feeding them before the onset of winter (see Chapter 8).

- **Carniolan *(A. m. carnica)*.** These bees are dark in color with broad gray bands. They originally hail from the mountains of Austria and Yugoslavia. This type exhibits a strong tendency to swarm. Carniolans maintain a small winter colony, which requires only small stores of food.

- **Caucasian *(A. m. caucasica)*.** Caucasian bees are mostly gray in color and are extremely adaptable to harsh weather conditions. They hail from the Caucasian Mountains near the Black Sea. They make extensive use of propolis to chink-up drafty openings, which can make quite a sticky challenge for the beekeeper. Caucasian bees also are prone to robbing honey, which can create a rather chaotic beeyard. They can also fall victim to Nosema disease, so be sure to medicate your Caucasian bees with Fumidil-B every spring and autumn (see Chapter 8).

- **Buckfast (hybrid).** The Buckfast bee was the creation of Brother Adam, a Benedictine monk at Buckfast Abby in the United Kingdom. Brother Adam earned a well-deserved reputation as one of the most knowledgeable bee breeders in the world. The precise heritage of the Buckfast bee seems to have been known only by Brother Adam — and sadly he died in 1996 at the impressive age of 98. He mixed the British bee with scores of bees from other races, seeking the perfect blend of gentleness, productivity, and disease resistance. The Buckfast bee's resulting characteristics have created quite a fan club of beekeepers from all around the world. The Buckfast bee excels at brood rearing, but exhibits a tendency, however, toward robbing and absconding from the hive (see Chapter 9 for information on how to prevent these bad habits).

- **Starline (hybrid).** This bee was derived as a hybrid strain of Italians and is the only commercially available hybrid race of Italians. It is regarded as productive at pollinating clover, so some people refer to the Starline as the *clover bee.*

- **Midnight (hybrid).** The double hybrid bee called *Midnight* is trademarked by York Bee Company in Gesup, Georgia. The Midnight bee makes heavy use of propolis, which can make inspecting a colony of Midnight bees a sticky challenge for the beekeeper.

- **Africanized (hybrid).** This bee is not commercially available, nor desirable to have. I mention it here because its presence has become a reality throughout South America, Mexico, and parts of the Southern U.S. The list of bee races is not complete without a nod to the so-called *Killer Bee.* This bee's aggressive behavior makes it difficult and even dangerous to manage. (See Chapter 9 for more on this type of bee).

The Russians are coming!
The Russians are coming!

The bee journals are all full of hoopla about the wondrous "Russian honey bee." This strain (recently available in the U.S.) is touted as being a good honey producer, mite and disease resistant, and gentle to deal with. Results emerging from informal test evaluations of this bee are encouraging. Apparently the Russian's heritage of prolonged winters and heavy mite populations enabled only the sturdiest of bees to survive. A true winter warrior! The Russian-American hybrids that now are offered by some dealers are certainly worth watching. Time will tell whether the Russian bees are all that their marketers claim them to be.

Generally speaking, the four characteristics that you should consider when picking out the bee strain that you want to raise are gentleness, productivity, disease tolerance, and how well the bees survive winters in the northern United States and Canada. Table 5-1 assigns the various types of bees listed above a rating from 1 to 3 in these 4 categories, with 1 being the most desirable and 3 the least desirable.

Table 5-1	Characteristics of Various Common Honey Bee Types			
Bee Type	*Gentleness*	*Productivity*	*Disease Tolerance*	*Wintering in Northern USA*
Italian	1	1	2	2
Carniolan	1	2	2	2
Caucasian	1	2	2	1
Buckfast (hybrid)	2	2	1	1
Starline (hybrid)	2	1	2	2
Midnight (hybrid)	2	2	2	1
Africanized	3	1	1	3

After all that's said and done, which kind of bee do I recommend you start with? Italian. No doubt about it. The number-one choice in the world is popular for a good reason: The Italian honey bee is gentle, productive, and does well in many different climates. It's a great bee for beginning beekeepers. Look no further in your first year.

At some point in years to come, you may want to try raising your own various races and hybrids of bees. Much is involved in breeding bees. It's a science that involves a good knowledge of biology, entomology, and genetics. A good book on the subject is *Queen Rearing and Bee Breeding* by Harry H. Laidlaw Jr. and Robert E Page Jr. (1997 Wicwas Press, ISBN: 1-878075-08-X).

Deciding How to Obtain Your Initial Bee Colony

You'll need some bees if you're going to be a beekeeper. But where do they come from? You have several different options when it comes to obtaining your bees. Some are good; others are not so good. This section describes these options and their benefits or drawbacks.

Ordering package bees

One of your best options and by far the most popular way to start a new hive is to order package bees. It's the choice that I most recommend. You can order bees by the pound from a reputable supplier. *Bee breeders* are found mostly in the southern states and will ship just about anywhere in the continental U.S.

A package of bees and a single queen are shipped in a small wooden box with two screened sides (see Figure 5-1). Packaged bees are sent via U.S. Mail or United Parcel Service (UPS). A package of bees is about the size of a large shoebox and includes a small screened cage for the queen (about the size of a matchbook) and a tin can of sugar syrup that serves to feed the bees during their journey. A three-pound package of bees contains about 11,000 bees, the ideal size for you to order. Order one package of bees for each hive that you plan to start.

Be sure to order a marked queen with the package. *Marked* means that a small colored dot has been painted on her thorax. This dot helps you spot the queen in your hive during inspections. It also confirms that the queen you see is the one that you installed (versus discovering an unmarked one that means your queen is gone and another has taken her place).

Be sure to pick a reputable dealer with a good track record for providing healthy and disease-free package bees (criteria for selecting a vendor is discussed in "Picking a Reputable Supplier" later in this chapter). When ordering, be sure to ask to see a copy of a certificate of health from the vendor's state apiary inspector. If the vendor refuses . . . be wary.

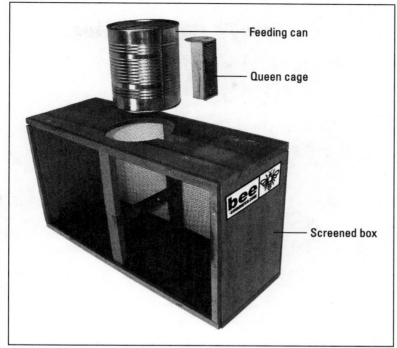

Feeding can

Queen cage

Screened box

Figure 5-1:
Package
bees are
shipped in
screened
boxes. Note
the feeding
can and
queen cage.

Buying a "nuc" colony

Another good option for the new beekeeper: Find a local beekeeper that can sell you a *nucleus (nuc)* colony of bees. A nuc consists of four to five frames of brood and bees, plus an actively laying queen. All you do is transfer the frames (bees and all) from the nuc box into your own hive. But finding someone who sells nucs isn't necessarily so easy, because few beekeepers have nucs for sale. After all, raising volumes of nucs for sale is a whole lot of work with little reward. But if you *can* find a local source, it's far less stressful for the bees (they don't have to go through the mail system). You can also be reasonably sure that the bees will do well in your geographic area. After all, it's already the place they call home! An added plus is that having a local supplier gives you a convenient place to go when you have beekeeping questions (your own neighborhood bee mentor). To find a supplier in your neck of the woods, check your yellow pages under "beekeeping," call your state's bee inspector, or ask members of a local beekeeping club or association.

A *nuc* or *nucleus* consists of a small wooden or cardboard hive (a "nuc box"; see Figure 5-2) with three to five frames of brood and bees, plus a young queen.

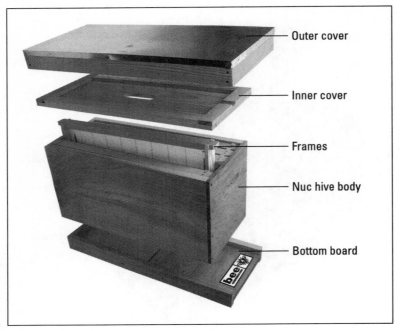

Outer cover

Inner cover

Frames

Nuc hive body

Bottom board

Figure 5-2:
A standard
nuc box.

Look for a reputable dealer with a good track record for providing healthy bees (free of disease). Ask whether the state bee inspector inspects the establishment annually. Request a copy of a certificate of health from the state. If you can find a reputable beekeeper with nucs, this is a convenient way to start a hive and quickly build up a strong colony.

Purchasing an established colony

You may find a local beekeeper who's willing to sell you a fully established colony of bees — hive, frames, bees, the whole works! This is fine and dandy, but more challenging than I recommend for a new beekeeper. First, you encounter many more bees to deal with. And the bees are mature and well established in their hive. They tend to be more protective of their hive than a newly established colony (you're more likely to get stung). Their sheer volume makes inspecting the hive a challenge. Furthermore, old equipment may be harder to manipulate (things tend to get glued together with propolis after the first season). More important, you also lose the opportunity to discover some of the subtleties of beekeeping that you can experience only when starting a hive from scratch: the building of new comb, introducing a new queen, and witnessing the development of a new colony.

Wait until you've gained more experience as a beekeeper. If you're determined, however, to select this option, make sure that you have your state's apiary inspector look at the colony before agreeing to buy it. You want to be 100 percent certain the colony is free of disease (for more information about honey bee diseases, see Chapter 10). After all, you wouldn't buy a used car without having a mechanic look at it first.

Capturing a wild swarm of bees

Here's an option where the price is right: Swarms are free. But I don't recommend this for the first-year beekeeper. Capturing a wild swarm is a bit tricky for someone who never has handled bees. And you never can be sure of the health, genetics, and temperament of a wild swarm. In some areas (mostly the southern U.S.) you face the possibility that the swarm you attempt to capture may be *Africanized* (see Chapter 9). My advice? Save this adventure for year two.

Picking a Reputable Supplier

By checking advertisements in bee journals and surfing the Internet you'll come up with a long list of bee suppliers (see Appendix A for a list of my favorite suppliers along with information on bee-related Web sites, journals, and organizations). But all vendors are not created equal. Here are some rules of thumb for picking a good vendor:

- **Be sure to pick a well-established vendor who has been breeding and selling bees for many years.** The beekeeping business is full of well-meaning amateurs who get in and out of breeding bees. They lack the experience that results in a responsible breeding program, which can result in problematic stocks of bees and lackluster customer service.

- **Look for a supplier with a reputation for consistently producing healthy bees and providing dependable shipping and good customer service.** Figure 5-3 shows a picture of a well-run commercial bee-breeding yard.

- **Ask if the establishment is inspected each year by the state's apiary inspector.** Request a copy of its health certificate. If owners refuse to comply, look elsewhere.

- **A reputable supplier replaces a package of bees that dies during shipment.** Ask potential suppliers about their replacement guarantee.

- **Be suspicious of suppliers who make extravagant claims.** Some walk a fine ethical line when they advertise that their bees are "mite or disease resistant." No such breed of bee exists. New beekeepers are easy prey for these charlatans. If it sounds too good to be true, it probably is. Look elsewhere.

✔ **Consult with representatives of regional bee associations.** Contact your state's apiary inspector or other bee association representatives. Find out whom they recommend as suppliers. Get them to share their experiences with you — good and bad.

✔ **Join a local bee club to get vendor recommendations from other members.** This also is a great way to find out more about beekeeping and latch onto a mentor. Many clubs have "new beekeeper" programs and workshops.

Figure 5-3:
A commercial beeyard.

Courtesy of Bee Culture Magazine

Deciding how many hives you want

Starting your adventure with *two* hives of bees offers certain advantages. Having two gives you a basis for comparison. It enables you to borrow frames from a stronger colony to supplement a weaker colony. In some ways two hives doubles the fun. You'll have more bees to pollinate your garden and more opportunities to witness what goes on within a colony. And, of course, you'll double your honey harvest! You can also double the rate of your learning curve. I suggest, however, that you begin with no more than two hives during your first year. More than two can be too much for the beginner to handle. Too many bees can be too time consuming and present too many new problems to digest before you really know the subtleties of beekeeping.

Deciding When to Place Your Order

When you're ordering packaged bees, you want to time your order so that you receive your bees as early in the spring as the weather allows. Doing so gives your colony time to build its numbers for the summer "honey flow" and means your bees are available for early pollination. Suppliers usually start shipping packaged bees early in April and continue through the end of May. Large commercial bee breeders shake bees into screened packages and ship hundreds of packages daily during this season (see Figure 5-4). After that, the weather simply is too hot for shipping packaged bees — they won't survive the trip during the scorching hot days of summer (most bees ship from the southern states). Local bee suppliers have nucs available in a similar time frame.

Bees are in limited supply and available on a *first-ordered, first-shipped basis*. Avoid disappointment. Place your order early. November is not too soon to order bees for the upcoming spring.

Figure 5-4: A commercial beekeeper shakes bees into boxes for shipping.

Courtesy of Bee Culture Magazine

The Day Your Girls Arrive

You may not know the exact day that your bees will arrive, but many suppliers at least let you know the approximate day they plan to *ship* your package bees.

About a week before the anticipated date of arrival, alert your local post office that you're expecting bees. Make sure that you provide the post office with your telephone number so you can be reached the moment your bees come in. In most communities the post office asks that you pick up your bees at the post office. Seldom are bees delivered live right to your door. Instruct the post office that the package needs to be kept in a cool, dark place until you arrive.

In all likelihood, you'll receive your "bees-have-arrived" call in the predawn hours — the instant they arrive at your local post office. Postal workers will, no doubt, be eager to get rid of that buzzing package! Please note, however, that this wake-up call is *not* the signal for you to start assembling your equipment. Plan ahead! Make sure everything is ready for your girls before they arrive.

Bringing home your bees

When the bees finally arrive, follow these steps in the order they are given:

1. **Inspect the package closely.**

   Make sure that your bees are alive. You may find some dead bees on the bottom of the package, but that is to be expected. If you find an inch or more of dead bees on the bottom of the package, however, fill out a form at the post office and call your vendor. He or she should replace your bees.

2. **Take your bees home right away (but don't put them in the hot, stuffy trunk of your car).**

   They'll be hot, tired, and thirsty from traveling.

3. **When your get home, spray the package liberally with cool water using a clean mister or spray bottle.**

4. **Place the package of bees in a cool place, such as your basement or garage, for an hour.**

5. **After the hour has passed, spray the package of bees with nonmedicated sugar syrup (see recipe that follows).**

   Don't *brush* syrup on the screen, because doing so literally brushes off many little bee feet in the process.

You must have a means for feeding your bees once they're in the hive. I strongly recommend using a good quality hive-top feeder. Alternatively you can use a feeding pail (See Chapter 4 for additional information on different kinds of feeders).

Recipe for sugar syrup

You'll need to feed your bees sugar syrup twice a year in spring and in autumn.

The early spring feeding stimulates activity in the hive and gets your colony up and running fast. It also may save lives if the bees' winter stores of honey have dropped dangerously low.

The colony will store the autumn sugar syrup feeding for use during the cold winter months.

In either case, feeding syrup is also a convenient way to administer some important medications.

If you purchased your bees from a reputable bee breeder, you won't need to medicate your bees during your first season. But you *will* want to feed them medicated syrup twice a year (spring and autumn) in your second and subsequent seasons.

- ✔ **Nonmedicated syrup:** Boil 2½ quarts of water on the stove. When it comes to a rolling boil, turn off the heat and add 5 pounds of white granulated sugar. Be sure you turn off the stove. If you continue boiling the sugar, it may caramelize, and that makes the bees sick. Stir until the sugar completely dissolves. The syrup must cool to room temperature before you can feed it to your bees.

- ✔ **Medicated syrup:** For medicated syrup, prepare the nonmedicated syrup recipe as above. Let it cool to room temperature. Mix 1 teaspoon of Fumidil-B in approximately a half a cup of cool water (the medication won't dissolve directly in the syrup). Fumidil-B protects your bees against nosema — a common bee illness (see Chapter 10 for more information on nosema and other diseases). Add the medication to the syrup and stir. You also can add two tablespoons of Honey B Healthy. This food supplement contains essential oils and has a number of beneficial qualities.

Putting Your Bees Into the Hive

The fun stuff comes next. Sure, you'll be nervous. But that's only because you're about to do something you've never done before. Take your time and enjoy the experience. You'll find that the bees are docile and cooperative. Read the instructions several times until you become familiar and comfortable with the steps. The photo illustrations in the color insert of this book provide a helpful visual cue — after all, a picture is worth a thousand words!

When I hived my first package of bees, I had my wife standing by with the instructions, reading them to me one step at a time. What teamwork!

Ideally, hive your bees in the late afternoon on the day that you pick them up, or the next afternoon. Pick a clear, mild day with little or no wind. If it's raining and cold, wait a day. If you absolutely must, you can wait several days to put them in the hive, but make certain that you spray them two or three times a day with sugar syrup while they're waiting to be introduced to their new home.

Whenever I hive a package of bees, I always invite friends and neighbors to witness the adventure. They provide great moral support, and it gives them a chance to see first-hand how gentle the bees actually are. Ask someone to bring a camera. You'll love having the photos for your scrapbook!

To hive your bees, refer to the photo sequence in the color section of this book ("Installing a Package of Bees") and follow these steps in the order they are given:

1. **Thirty minutes before hiving, spray your bees rather heavily with nonmedicated sugar syrup (see photo number 1 in the color section, "Installing a Package of Bees").**

   But don't drown them with syrup. Use common sense and they'll be fine.

2. **Using your hive tool, pry the wood cover off the package (see photo number 2 in the color section, "Installing a Package of Bees").**

   Pull the nails or staples out of the cover, and keep the wood cover handy.

3. **Jar the package down sharply on its bottom so that your bees fall to the bottom of the package (see photo number 3 in the color section, "Installing a Package of Bees").**

   It doesn't hurt them! Remove the can of syrup from the package (see photo number 4 in the color section, "Installing a Package of Bees") and the queen cage, and loosely replace the wood cover (without the staples).

4. Examine the queen cage. See the queen?

She's in there with a few attendants. Is she okay? In rare cases, she may have died in transit. If that's the case, go ahead with the installation as if everything were okay. But call your supplier to order a replacement queen (there should be no charge). Your colony will be fine while you wait for your replacement queen.

5. Slide the metal disc on the queen cage to the side slowly.

Remove the cork at one end of the cage so that you can see the white candy in the hole. If the candy is present, remove the disc completely. If the candy is missing, you can plug the hole with a small piece of marshmallow. If your package comes with a strip of *Apistan* (designed to control mites during shipment), remove it from the back of the queen cage.

6. Out of two small frame nails bent at right angles, fashion a hanging bracket for the queen cage (see photo number 5 in the color section, "Installing a Package of Bees").

7. Spray your bees again, and jar the package down so the bees drop to the bottom.

8. Prepare the hive by removing five of the frames, but keep them nearby.

Remember that at this point in time you're using only the lower deep hive body for your bees. Now hang the queen cage (candy side up) between the center-most frame and the next frame facing toward the center (see photo number 6 in the color section, "Installing a Package of Bees"). The screen side of the cage needs to face toward the center of the hive.

9. Spray your bees liberally with syrup one last time.

Jar the package down. Toss away the wood cover and then pour (and shake) approximately half of the bees directly above the hanging queen cage (see photos 7 and 8 in the color section, "Installing a Package of Bees"). Pour (and shake) the remaining bees into the open area created by the missing five frames.

10. When the bees disperse a bit, gently replace four of the five frames (see photo number 9 in the color section, "Installing a Package of Bees").

Do this gingerly so you don't crush any bees. If the pile of bees is too deep, use your hand (with gloves on) to disperse the bees.

11. Place the inner cover on the hive.

If you're using a hive-top feeder, it is placed in direct contact with the bees without the inner cover in between, so skip this step and go to step 12. The inner cover is used only when a jar or pail is used for feeding. The outer cover is placed on top of the hive-top feeder.

12. **Place the hive-top feeder on top of the hive (see photo number 10 in the color section, "Installing a Package of Bees").**

   Alternatively, invert a one-gallon feeding pail above the oval hole in the inner cover; add a second deep super on top of the inner cover; and fill the cavity around the jar with crumpled newspaper for insulation (see photo number 11 in the color section, "Installing a Package of Bees").

13. **Plug the inner cover's half-moon ventilation hole with a clump of grass.**

   You want to close off this entrance until the bees become established in their new home.

14. **Now place the outer cover on top of hive. You're almost done.**

15. **Insert your entrance reducer, leaving a one-finger opening for the bees to defend (see photo number 12 in the color section, "Installing a Package of Bees").**

   Leave the opening in this manner until the bees build up their numbers and can defend a larger hive entrance against intruders. This takes about 4 weeks. If an entrance reducer isn't used, use grass to close up all but an inch or two of the entrance.

   Place the entrance reducer so that the openings face "up." Doing so allows the bees to climb up over any dead bees that might otherwise clog the small entrance.

You're done! Take a breath, and leave everything alone for a week. No peeking! The bees may kill the queen if they're disturbed before five days have elapsed after her introduction.

Use this first week to get to know your bees. Take a chair out to the hive and sit to the side of the entrance — about two to three feet from the hive (within reading distance). Watch the bees as they fly in and out of the hive. Some of the workers will return to the hive with pollen on their hind legs. Other bees will be fanning at the entrance ventilating the hive or releasing a sweet pheromone into the air. This scent is unique to this hive and helps guide their foraging sisters back to their home. Can you spot the guard bees at the entrance? They're the ones alertly checking each bee as she returns to the hive. Do you see any drones? They are the male bees of the colony and are slightly larger and more barrel-shaped than the female worker bees. The loud, deep sound of their buzzing often distinguishes them from their sisters.

Congratulations! You're now officially a beekeeper. You've launched a wonderful new hobby that can give you a lifetime of enjoyment.

Knowing when and how to use the entrance reducer

The entrance reducer is used for two primary reasons:

- To regulate the hive's temperature

- To restrict the opening so that a new or weak colony can better defend the colony.

That being the case, here are some guidelines:

- For a newly hived colony, leave the entrance reducer in place (utilizing its smallest opening) until approximately six weeks after you hived your package of bees. Chances are you can then position the entrance reducer so that its next largest opening is utilized. After about eight weeks, you can remove the entrance reducer completely. By that time the colony should be strong enough to defend itself — and the weather should be warm enough to fully open the entrance.

- For an established colony, use the entrance reducer during long periods of cold weather (less than 40° F; 4° C). It helps prevent heat from escaping from the hive. I prefer not to use the smallest opening, as I find it too restrictive — bees that die from attrition can clog the small opening. As a general "rule of thumb," remove the entrance reducer completely when daytime temperatures are above 60° F; 15° C.

Part III
Time for a Peek

The 5th Wave
By Rich Tennant

"Oh really, Gerald! You're removing honey from the hives, not plutonium!"

In this part . . .

This is where you get up-close and personal with your honey bees. You learn the best and safest way to inspect and enjoy your bees, as well as maintain your colony year-round. I also share useful tips and techniques that help you develop good habits right from the start.

Chapter 6

Opening Your Hive

· ·

In This Chapter

▶ Knowing when and how often to visit your bees

▶ Finding out how to light your smoker

▶ Deciding what to wear

▶ Approaching the hive

▶ Opening the hive

· ·

*T*his is the moment you, as a new beekeeper, have been waiting for — that exhilarating experience when you take your first peek into the hive. You likely have a touch of fear, tempered by a sweeping wave of curiosity.

Put those fears aside. You'll soon discover visiting with your bees is an intoxicating experience that you eagerly look forward to. What you're about to see is simply fascinating. It's also one of the more tranquil and calming experiences that you can imagine: The warmth of the sun; the sweet smell of pollen, wax, and honey; the soothing hum of the hive. You're at one with nature. Your new friendship with your bees will reward you for many years to come.

The habits you develop in the beginning are likely to stick with you. So developing good habits early on is important. By learning the safe and proper way to inspect your hive and following suggested steps religiously in the beginning, you'll minimize any risks of injuring or antagonizing your bees. The techniques become second nature in no time. Down the road, you may find variations on the suggested methods that suit you better. That's okay. Just relax, move calmly, use good judgment, and enjoy the miracle of beekeeping.

Visiting Your Bees at the Best Time

Ideally, open your hive on a nice sunny day. Between 10 a.m. and 5 p.m. is best. Under those conditions, thousands of busy worker bees are out in the field. Avoid cold, windy, or rainy weather, because that's when the entire colony is at home. With everyone in the hive, you'll probably find too many bees to deal with. In addition, they tend to be crankier when they can't get out of the house! You know how that is.

Setting an Inspection Schedule

For the new beekeeper, once a week isn't too often to visit the bees. Use these frequent opportunities to find out more about the bees and their life cycles. Your first season is a time of discovery. You'll begin recognizing what's normal and what's not. You'll also become increasingly comfortable with manipulating the frames and working with the bees. So much so that it soon becomes second nature and a quick peek at the entrance or under the lid is all that's needed to assure you that all is well. Beekeeping is as much an art as it is a science. Practice makes perfect.

Once you begin getting the hang of it, you needn't conduct more than six to eight thorough inspections a year: Three or four visits in the early spring, one or two during the summer, and a couple of inspections at the end of the season are all that are necessary. It's better not to disturb your bees too often.

Every day that you smoke the bees, open their hive, and pull it apart sets their productivity back a bit. It takes a day or two for life in the colony to return to normal. So if harvesting lots of honey is your objective, limit your inspections to once every few weeks.

This schedule doesn't apply to your first year — when you need to gain greater experience by visiting the hive often.

Preparing to Visit Your Hive

The weekend has rolled around and the weather's great (mild, sunny, and not much wind), so you've decided that you're going to pay the girls a visit. It's time to see what's going on in the hive. But you can't just dash out and tear the top off the hive. You have to get yourself ready for this special occasion. What will you wear? How will you approach them? What in the world do you do with all this new equipment?

In the upcoming section, I'll take you through the details of each step. You may want to read this chapter and the next one word for word. You may even want to read it a few times before having your first "close encounter." You may also want to take the book along on your first inspection, just in case you need some quick moral support. Better yet, coerce a friend or family member to go with you. That's what I did the first time. I had my wife reading from a book and prompting me through each step of the way. At the time we didn't have an extra veil for her, so she hollered instructions at me from a safe distance.

Making "non-scents" a part of personal hygiene

Forgive me for being personal. But you need to know that bees don't react well to bad body odor. So, please don't inspect your bees when you're all sweated up after a morning jog. Take a shower first. Brush your teeth. On the other hand, don't try to smell too good, either!

Avoid using colognes, perfumes, or scented hairsprays. Sweet smells can attract more attention from the bees than you want.

Be sure to remove your leather watchband before visiting your bees. They don't like the smell of leather or wool. Removing any rings from your fingers also is a good idea. It isn't that bees don't like pretty jewelry. But in the rare event that you take a sting on your hand, you don't want your fingers swelling up when you're wearing a decidedly nonexpandable ring.

Getting dressed up and ready to go

Always wear your veil when you're inspecting your hive. Doing so keeps the bees away from your face and prevents them from getting tangled in your hair. For a discussion of the types of veils that are available, see Chapter 4.

If a bee ever gets under your veil, try not to panic. It isn't that big a deal. She's unlikely to sting you unless you squeeze her. Simply walk away from the hive and slip off your veil. Don't remove your veil at the hive, and don't thrash around screaming and yelling. Doing so only upsets the bees, and the neighbors will think you've gone wacky.

New beekeepers need to wear a long-sleeved shirt. Light colors and smooth fabrics (like cotton) are best, because bees don't like dark colors, or the smell of wool or leather (material made from animals). Using elastic, Velcro straps, or rubber bands around the cuff of each pant leg and sleeve keeps clothing bee-tight, unless, of course, you think you might actually like having curious bees traveling up inside your trousers.

You can use gloves if you feel you absolutely must (see Chapter 4 for more information about gloves). But I encourage you not to develop that habit. Gloves are bulky. They impair your sense of touch and make your movements clumsy. When you're working with new colonies and early spring colonies, gloves aren't even necessary. These small, young, and gentle colonies are a delight to work with. Save your gloves for unfavorable weather, moving colonies around, or for use during the late summer and honey harvesting time (when the colony's population is large and bees tend to be more defensive). But at all other times, I recommend that you leave the gloves at home. Trust me.

Colonies can be handled with far more dexterity and fewer injuries to bees (and you!) when you don't use gloves at all. Less injury to bees means a more docile colony.

Lighting your smoker

The smoker is the beekeeper's best friend. Yet for many, keeping a smoker lit can be the trickiest part of beekeeping. It doesn't have to be. What you're trying to achieve is enough thick, cool smoke to last throughout your inspection. You certainly don't want your smoker to poop out as soon as you've opened the hive.

Begin with a loosely crumpled piece of newspaper about the size of a tennis ball. Light the paper and place it in the bottom of the smoker. Nest it in place using your hive tool. Gently squeeze the bellows a few times until you're sure that the paper is burning with a flame.

Add dry matchstick-size kindling, pumping the bellows as you do. As it ignites (you'll hear it crackling), slowly add increasingly thicker kindling. Ultimately, the fattest of your twigs will be about as thick as your thumb. None of the kindling need be more than four or five inches in length. The kindling needs to fill three quarters of the smoker, and must be thoroughly packed from side to side. Using your hive tool, occasionally stoke the fire. Keep pumping. When your kindling has been burning for about 10 minutes, and embers are glowing, it's time to add the real fuel.

Use a fuel that burns slowly and gives off lots of smoke. I am partial to dry wood chips or hemp baling twine. But burlap, dry leaves, and even pine needles do nicely. You can also purchase *smoker fuel* (usually cartridges of compacted raw cotton fibers, or nuggets of wood) from beekeeping supply stores. It works well, too. The bees really don't care what you use — but avoid using anything synthetic or potentially toxic. Figure 6-1 shows a smoker and various kinds of starters and fuels.

Keep a box of kindling and fuel with your other beekeeping equipment. Having this readily available saves time on the days that you plan to visit the hive.

Pack the smoker right to the top with your preferred fuel, as you continue to gently pump the bellows. When billows of thick, cool, and white smoke emerge, close the top. Pump the bellows a few more times. Use a long, slow pumping method when working the bellows, rather than short, quick puffs. Doing so produces more and thicker smoke than short puffs (see Figure 6-2).

Congratulations! You're now ready to approach the hive. Your smoker should remain lit for many hours.

Figure 6-1:
A smoker with all the "ingredients" is ready to load with paper, kindling of various sizes, and some hemp twine.

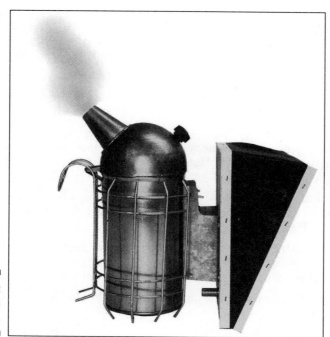

Figure 6-2:
A smoking smoker.

Keeping your smoker clean

A good question that I'm frequently asked is: "My smoker is all gummed up and needs a good cleaning. How do I clean it?"

After a season or two, the inside of a smoker can become thickly coated with black, gummy tar. I've found the best way to clean it is by burning the tar out of it — literally. Like a self-cleaning oven, you need a great amount of heat. I've had success using a small propane blowtorch. You can purchase one at any hardware store. Just apply the flame to the black tar coating the inside of your smoker. Keep blasting away. Soon the tar ignites, glows a fiery orange, and then turns to a powdery ash. Turn off the blowtorch. Once the metal smoker cools, you can easily knock the ash out of the smoker. Clean as a whistle!

Opening the Hive

You're all suited up and you have your smoker and hive tool. Perfect! Be sure to bring along an old towel (I'll explain why later in the "Removing the hive-top feeder" section). So now the moment of truth has arrived. Approach your hive from the side or rear. Avoid walking right in front of it, because the bees shooting out the entrance will collide with you. As you approach the hive, take a moment to observe the bees and then ask yourself, "In what direction are they leaving the hive?" Usually it's straight ahead, but, if they're darting to the left or right, approach the hive from the opposite side. Follow these steps to open the hive:

1. **Standing at the side and with your smoker 2 or 3 feet from the entrance, blow several puffs of thick, cool smoke into the hive's entrance (see Figure 6-3).**

   Four good puffs of smoke should do fine. Use good judgment. Don't over-smoke them. You're not trying to asphyxiate the bees; you simply want to let the guard bees know you're there.

2. **Still standing at the side of the hive? Good. Now lift one long edge of the outer cover an inch or so, and blow a few puffs of smoke into the hive (see Figure 6-4).**

   Ease the top back down and wait 30 seconds or so. Doing so gives the smoke time to work its way down into the hive. These puffs are for the benefit of any guard bees at the top of the hive.

3. **Put your smoker down and, using both hands, slowly remove the outer cover.**

   Lift it straight up and off the hive. Set the cover upside down on the ground (with the flat metal top resting on the ground, and its underside facing skyward).

Figure 6-3:
Approach the hive from the side and blow a few puffs of smoke into the entrance to calm the guard bees.

Figure 6-4:
A little smoke under the hive cover calms any of the colony's guard bees that may be upstairs.

What does the smoke do?

Smoke calms bees and prevents them from turning aggressive during inspections. You may ask, "Why?" One explanation is that it tricks bees into thinking there's a fire. In nature bees make their homes in hollow trees. So a forest fire would be a devastating event. Smelling the smoke, the bees fan furiously to keep the hive cool. They also begin collecting their most precious commodity — honey, engorging their *honey stomachs* with it in the event they must abandon ship and move to a new and safer home. With all the commotion, they become quite oblivious to the beekeeper. And when the inspection is complete and the crisis passes, the bees return the honey to the comb. That way, nothing is lost.

Another explanation is that the smoke masks the alarm pheromones given off by worker bees when the hive is opened. Ordinarily, these alarm pheromones trigger defensive action on the part of the colony. But the smoke confounds the bees' ability to communicate danger. In any event, smoking the bees really works. Don't even think about opening a hive without first smoking it. It's a tempting shortcut that you'll try only once.

Your next step depends on whether you're still feeding your bees at the time of the inspection. If no hive-top feeder is on the colony, skip ahead to the section on "Removing the inner cover."

Removing the hive-top feeder

If you're using a hive-top feeder, you'll need to remove it before inspecting your hive. To do so, follow these steps:

1. **With your smoker, puff some smoke through the screened access, and down into the hive (see Figure 6-5).**

2. **Hive parts often stick together, so use the flat end of your hive tool to gently pry the feeder from the hive body (see Figure 6-6).**

   Do this slowly, being careful not to pop the parts apart with a loud "snap." That only alarms the bees.

   Here's a useful trick. Use one hand to gently press down on the feeder, while prying the feeder loose with the hive tool in your other hand. This *counterbalance* of effort minimizes the possibility of the two parts suddenly popping apart with a loud "snap."

3. **Loosen one side of the feeder and then walk around and loosen the other side.**

Figure 6-5:
If you're using a hive-top feeder, apply some smoke through the screened access to reach the bees down below.

Figure 6-6:
Use your hive tool as a lever to ease apart hive parts.

4. **Blow a few puffs of smoke into the crack created by your hive tool as you pry loose the feeder.**

5. **Wait 30 seconds and completely remove the hive-top feeder.**

   Be careful not to spill any syrup. Set the feeder down on the outer cover that now is on the ground.

Positioning the feeder at right angles to the cover when you set it down, results in only two points of contact and makes it less likely that you'll crush any bees that remain on the underside of the feeder. Always be gentle with them, and they'll always be gentle with you!

Remember the old towel I talked about earlier in this chapter? This is where it comes in. If syrup remains in the feeder, completely cover it with the towel (alternatively you can use a small plank of plywood or a scrap of carpeting). Syrup left in the open attracts the bees — big time! You don't want to set off *robbing*. That's a nasty situation where bees go into a wild frenzy after finding free sweets (see Chapter 9). Open containers of syrup (or honey for that matter) also can attract bees from other colonies. All the gorging bees wind up whipped into such a lather that they begin robbing honey from your hive. War breaks out, and hundreds or even thousands of bees can be killed by the robbing tribe. Enough said! A good rule of thumb: *Never* leave syrup in the wide open. Keep it out of reach!

Removing the inner cover

If you're *not* using a hive-top feeder, you'll need to remove the inner cover (an inner cover is always used *unless* a top feeder is on the hive). Removing the inner cover is much like removing the top feeder. Follow these steps:

1. **Puff smoke through the oval hole and down into the hive.**

2. **Using the flat end of your hive tool, gently release the inner cover from the hive body (see Figure 6-7).**

   Loosen one side and then walk around and loosen the other side. Pry slowly, being careful not to pop the parts apart with a loud "crack."

3. **Blow a few puffs of smoke into the crack created by your hive tool as you pry up the inner cover.**

4. **Wait 30 seconds and then completely remove the inner cover.**

   Set it down on the outer cover that's now on the ground, or simply lean it up against a corner of your hive. Careful! Don't crush any bees that may still be on the inner cover.

Figure 6-7:
Direct some
smoke
through
the hole in
the inner
cover —
wait half
a minute
and then
remove it.

The Hive's Open! Now What?

Whew! With the feeder or inner cover removed, the hive is officially open.
Relax and take a deep breath. You should see lots of beautiful bees! Here's
what to do next:

1. **Time for the smoker again.**

   From one or two feet away, and standing at the rear of the hive, blow
   several puffs of cool smoke between the frames and down into the hives
   (see Figure 6-8). Pumping the bellows in long, slow puffs, rather than
   short, quick ones, make sure that the breeze isn't preventing smoke from
   going into the spaces between the frames. Watch the bees. Many of them
   will retreat down into the hive.

2. **Now you can begin your inspection (see Chapter 7).**

   Although you have much to do, you don't want to keep the hive open for
   more than 10 to 15 minutes (even less if the weather is cooler than 55
   degrees F). But don't rush at the expense of being careful! Clumsiness
   results in injury to bees, and that can lead to stings. Be gentle with the
   ladies!

In Chapter 7, I'll explain exactly what you should look for when the hive's
wide open.

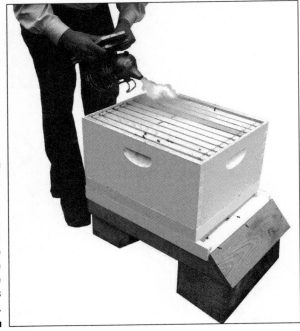

Figure 6-8:
Blowing
smoke
between the
frames and
down into
the hive
drives the
bees
downward.

Chapter 7

Knowing What to Look For

- -

- -

*P*eering out through your veil with your cuffs strapped shut and your smoker lit, you've opened your hive and now see that it's bustling with bees. But what exactly are you looking for?

Understanding when to look and what to look for indicates the difference between being a "beekeeper" and a "bee*haver*." Anyone can have a hive of bees, but your goal as a beekeeper is to help these little creatures along. Understand their needs. Learn to anticipate. Give them the room they need before they actually need it. Give them comb in which to store honey before the nectar starts to flow. Medicate and feed them before an emergency strikes. Get them ready for winter before the weather turns cold. In return, your bees will reward you with many years of enjoyment and copious crops of sweet golden honey.

Exploring Basic Inspection Techniques

The approach for inspecting your hive doesn't vary much from one visit to another. You always follow certain procedures, and you always look for certain things. After a few visits to the hive, the mechanics of all this become second nature, and you can concentrate on enjoying the miraculous discoveries that await you. In this section I'll give you some pointers that make each inspection easy.

Removing the first frame

Always begin your inspection of the hive by removing the *first frame* or *wall frame*. That's the frame closest to the outer wall. Which wall? It doesn't matter. Pick a side of the hive to work from, and that determines your first frame. Here's how to proceed:

1. **Insert the curved end of your hive tool between the first and second frames, near one end of the frame's top bar (see Figure 7-1).**

2. **Twist the tool to separate the frames from each other.**

   Your hand moves toward the center of the hive — not the end.

3. **Repeat this motion at the opposite end of the top bar.**

   The first frame should now be separated from the second frame.

4. **Using both hands, pick up the first frame by the end bars (see Figure 7-2).**

   Gently push any bees out of the way as you get ahold of the end bars. With the frame in both hands *slowly* lift it straight up and out of the hive. Be careful not to roll or crush bees as you lift the frame. Easy does it!

You never should put your fingers on a frame without first noting where the bees are, because you don't want to crush any bees, and you don't want to get stung. Bees can be easily and safely coaxed away by gently pushing them aside with your fingers.

Figure 7-1:
Use your hive tool to pry the wall frame loose before removing it.

Figure 7-2:
Carefully lift out the first frame and set it aside. Now you have room to manipulate the other frames.

Now that you've removed the first frame, gently rest it on the ground, leaning it vertically up against the hive. It's okay if bees are on it. They'll be fine. Alternatively, if you have a frame rest (a handy accessory available at some beekeeping supply stores) use it to temporarily store the frame.

This is a basic and important first step every time you inspect a colony. The removal of this frame gives you a wide-open empty space in the hive for better manipulating the remaining frames without squashing any bees. Always be sure to remove the wall frame from the hive before attempting to remove any other frames.

Working your way through the hive

Using your hive tool, loosen frame two and move it into the open slot where frame one used to be. That gives you enough room to remove *this* frame without the risk of injuring any bees. When you're done looking at this frame, return it to the hive, close to (but not touching) the wall. Do *not* put this frame on the ground.

You work your way through all ten frames in this manner — moving the next frame to be inspected into the open slot. When you're done looking at a frame, always return it snugly against the frame previously inspected. Use your eyes to monitor progress as the frames are slowly nudged together.

Be careful not to crush any bees when pushing the frames together. One of those bees may be the queen! Look down between the frames to make sure the coast is clear before slowly pushing the frames together. If bees are on the end bars and at risk of being crushed, you can use the flat end of your hive tool to gently coax them to move along. A single puff of smoke also urges them to move out of the way.

Holding up frames for inspection

Discovering the proper way to hold and inspect an individual frame is important. Be sure to stand with your back to the sun, with the light shining over your shoulder and onto the frame (Figure 7-3). The sun illuminates details deep in the cells and helps you to better see eggs and small larvae. Here's an easy way to inspect both sides of the frame (Figure 7-4 illustrates the following steps):

1. **Hold the frame firmly by the tabs at either end of the top bar.**

   Get a good grip. The last thing you want to do is drop a frame covered with bees. Their retaliation for your clumsiness will be swift and, no doubt, memorable.

2. **Turn the frame vertically.**

3. **Then turn the frame like a page of a book.**

4. **Now smoothly return it to the horizontal position and you'll be viewing the opposite side of the frame.**

Figure 7-3: Hold frames firmly with the light source coming over your shoulder and onto the frame.

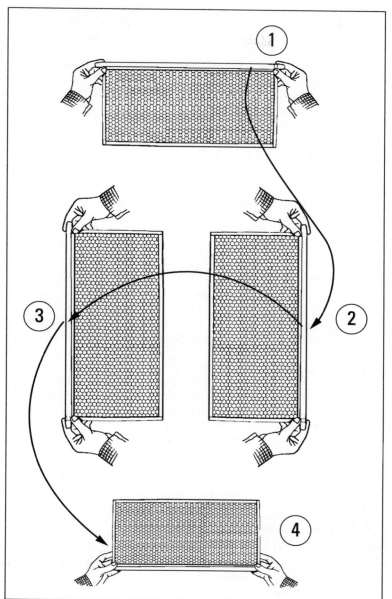

When inspecting frames, all your movements must be slow and deliberate. Change hand positions sparingly. Sliding your fingers across the frames as you reposition your hands is better than lifting your fingers and setting them down again, because you may land on a bee. As you turn the frame, you want to avoid any sudden and unnecessary centrifugal force that can disturb the bees.

Knowing when it's time for more smoke

A few minutes into your inspection, you may notice that the bees all have lined up between the top bars like racehorses at the starting gate. Their little heads are all in a row between the frames. Kind of cute, aren't they? They're watching you. That's your signal to give the girls a few more puffs of smoke to disperse them again so that you can continue with your inspection.

Understanding what to always look for

Each time that you visit your hive, be aware of the things that you always must look for. Virtually all inspections are to determine the health and productivity of the colony. The specifics of what you're looking for vary somewhat, depending upon the time of year. But some universal rules-of-the-road apply to every hive visit.

Checking for your queen

Every time that you visit your hive you're looking for indications that the queen is alive and well and laying eggs. If you actually see her, that's great and reassuring! But finding the queen becomes increasingly difficult as the colony becomes larger and more crowded. So how can you tell whether she's there?

Rather than spending all that time trying to see the queen, looking for *eggs* is better. Although they're tiny, finding the eggs is much easier than locating a single queen in a hive of 60,000 bees. Look for eggs on a bright, sunny day. Hold the comb at a slight angle and with the sun shining over your shoulder. This illuminates the deep recesses of the cells. The eggs are translucent white, resembling a miniscule grain of rice (see figures 2-9 and 2-10 in Chapter 2).

An inexpensive pair of reading glasses can help you spot the eggs — even if you don't normally need them. When you see eggs, you can be sure a queen is in the hive — or at least that she was there within the last two days.

Storing food; raising brood

Each deep frame of comb contains about 7,000 cells (3,500 on each side). Honeybees use these cells for storing food and raising brood. When you inspect your colony, noting what's going on in those cells is important because it helps you judge the performance and health of your bees. Ask yourself: Is there ample pollen and nectar? Are there lots of eggs and brood? Does the condition of the wax cappings over the brood look *normal* — or are the capping perforated and sunken in (see Chapter 9 for tips on recognizing unhealthy situations)?

Inspecting the brood pattern

Examining brood pattern is an important part of your inspections. A tight, compact brood pattern is indicative of a good, healthy queen (see "Things to Look For" in this book's color section). Conversely, a spotty brood pattern (many empty cells with only occasional cells of eggs, larvae, or capped brood) is an indication that you have an old or sick queen and may need to replace her. How does the capped brood look? These are cells that the bees have capped with a tan wax. The tan cappings are porous and enable the developing larvae within to breathe. The cappings should be smooth and slightly convex. Sunken-in (concave) or perforated cappings indicate a problem. See Chapter 10 for more information about how to recognize the telltale signs of brood disease.

Capped brood refers to larvae cells that have been capped with a wax cover, enabling the larvae to spin cocoons within and turn into pupae.

Recognizing foodstuffs

Learn to identify the different materials collected by your bees and stored in the cells. They'll pack pollen in some of the cells. Pollen comes in many different colors: orange, yellow, brown, gray, blue, and so on. The image on the cover of this book is a dramatic illustration of just how varied the colors of pollen can be. You'll also see cells with something "wet" in them. It may be nectar. Or it may be water. Bees use large amounts of water to cool the hive during hot weather. Other cells contain capped and cured honey. These cappings usually are bright white and airtight (versus the tan, porous cappings covering brood cells).

Replacing frames

After you've inspected your last frame, nine frames should be in the hive and one leaning against it or hanging on the frame rest (the first frame you removed). Putting the first frame back in the hive means:

1. **Slowly pushing the nine frames that are in the hive as a single unit toward the opposite wall of the hive.**

   That puts them back where they were when you started your inspection. Pushing them as a single unit keeps them snugly together and avoids crushing bees.

2. **You're now left with the open slot from which the first frame was removed.**

   Smoke the bees one last time to drive them down into the hive.

3. **Picking up the frame that's outside the hive.**

   Are bees still on it? If so, with a downward thrust, sharply knock one corner of the frame on the bottom board at the hive's entrance. The bees fall off the frame and begin walking into the entrance to the hive. With no bees remaining on your first frame, you can easily return it to the hive without the risk of crushing them.

4. **Easing the wall frame into the empty slot.**

   Slowly, please! Make certain that all ten frames fit snugly together. Using your hive tool as a wedge, adjust the ten-frame unit so that the space between the frames and the two outer walls is equal.

Closing the hive

You're almost finished. Follow these steps to close the hive:

1. **If you're using a hive-top feeder, put it back in place immediately on top of the hive body.**

   Add more sugar syrup if the pantry is getting low. Now go to Step 4.

2. **If you're not using a hive-top feeder, replacing the inner cover comes next.**

   First remove any bees from the inner cover. Use a downward thrust and sharply knock one corner of the inner cover on the bottom board at the hive's entrance.

3. **Place the inner cover back on the hive by sliding it in position from the rear of the hive so that you don't crush any bees.**

   Very slowly slide it into place, and any bees along the top bars or on the edges of the hive will be pushed gently out of the way. Kind of like a bulldozer!

   Note that the notched ventilation hole is positioned upward and toward the front of the hive. This notched opening allows air to circulate and gives bees a top floor entrance to the hive.

4. **Replace the outer cover (the final step).**

   Make sure the outer cover is free of any bees. From the rear of the hive, slide it along the inner cover, again, gently pushing any bees out of the way (the bulldozer technique). Ease it into place, and adjust it so that it sits firmly and level on the inner cover.

Make sure that the ventilation notch on the outer cover isn't blocked. From the rear of the hive, shove the outer cover toward the front of the hive. Doing so opens the notched ventilation hole in the inner cover and gives the bees airflow and an alternate entrance.

Congratulations! The bees once again are snugly in their home.

The Bountiful Bee

Honey bees gather pollen and store it in pollen baskets located on their hind legs. Pollen is used by bees as a protein food.

These two workers share nectar. This form of socialization helps bees communicate the type of food sources available for foraging.

Foraging worker bees gather water from a leaky water spigot. The bees use water to cool the hive and dilute honey.

Installing a Package of Bees

All photos on this spread courtesy of Howland Blackiston

1. Thirty minutes before hiving, spray your packaged bees heavily with nonmedicated sugar syrup.

3. Jar the package down sharply on its bottom so that your bees fall to the bottom of the package.

2. Use your hive tool to pry the wood cover off the package.

4. Remove the feeding can and the queen cage. Loosely replace the wood cover.

5. Bend two small nails at right angles to fashion a hanging bracket for the queen cage. The candy plug must be facing up (top of the cage).

6. Prepare the hive for the bees by removing 5 of the frames. Hang the queen cage between two frames that are near the center of the hive.

7. Spray your packaged bees with syrup one last time. Jar the package down. Toss away the wood cover. Then pour and shake the bees directly over the hanging queen cage and into the open area created by the missing 5 frames.

8. Congratulations! The bees are in the hive.

9. Gently replace 4 of the 5 frames.

10. Put your hive top feeder directly on top of the deep hive body. The inner cover is not used when the hive top feeder is in place.

11. Fill the feeder with medicated syrup and put the outer cover on top of the feeder.

12. Place the entrance reducer so that the smallest opening is in use. Take a breath and leave everything alone for a week. No peeking.

Things to Look For

A beautiful frame of capped brood. Note the tight brood pattern with capped honey in the upper third of the frame.

These two worker bees are forcing a drone (male bee) to leave the colony. All drones are banished from the hive as winter approaches.

Guard bees will fight to the death to keep unwelcome insects, such as this bumblebee and the yellow jackets, out of the hive.

This nice swarm is ready to be put into a new hive. Honey bees are remarkably gentle during swarming and can be handled with relative ease.

Getting Close to Your Bees

Looking down into these cells, you see tiny rice-shaped eggs and young larvae. Spotting eggs is a good way to determine you have a queen.

These young worker bees are just emerging from their cells. Note the soft downy fur on these youngsters.

These two peanut-shaped cells contain developing queens.

Close-up of a worker bee's head. Note her extended tongue that is used like a straw to suck up water, nectar, and honey.

You can really see the huge wrap-around eyes on this young drone that is emerging from his cell.

Spotting Problems

The very spotty brood pattern on this frame is reason to suspect American foulbrood. The cappings are suspiciously sunken and perforated.

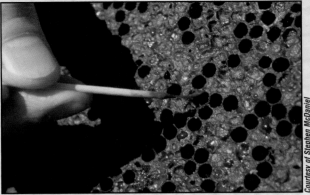

Check for American foulbrood by inserting a toothpick or matchstick into a suspect capped cell. If you note a ropy, gooey mass as you slowly withdraw the stick, you have good reason to believe your colony has American foulbrood.

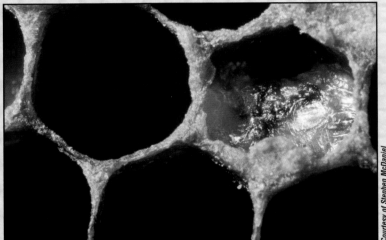

This scaly mass at the bottom of this cell is all that remains of this larva infected with European foulbrood.

The wax moth can create trouble for honey bees. Be sure to fumigate your honey supers with paradichlorobenzene (PDB) crystals before storing frames for the winter.

Wax moths can destroy honeycomb leaving a hopeless mess for the beekeeper. This frame is a goner.

These hard, chalky, mummified pupae at this hive's entrance indicate chalkbrood disease.

Courtesy of Stephen McDaniel

Courtesy of Stephen McDaniel

Close-up showing mummified chalkbrood larvae.

Courtesy of Wellmark International

Varroa mites are visible to the naked eye. Note the varroa on this worker bee.

Products of the Honey Bee

Honey comes in as many colors and flavors as there are flowers.

Courtesy of National Honey Board

Decide whether you want to harvest extracted (liquid) honey or chunks of natural comb honey. Better yet, why not produce both?

Courtesy of National Honey Board

Courtesy of Howland Blackiston

Honey's not the only food you can get from your bees. Try your hand at brewing mead (honey wine). Delicious!

Your New Colony's First Eight Weeks

For the newly hived colony, some specific beekeeping tasks are unique to the first few weeks of your first season. When you do any inspection, the general method for smoking, opening, and removing the frames is identical to the method given in the "Exploring Basic Inspection Techniques" section earlier in this chapter.

Looking in: A week after hiving your bees

After putting your package of bees in the hive, you'll be impatient to look inside to see what's happening. Resist the temptation! You must wait one full week before opening the hive. The colony needs this first uninterrupted week for accepting its new queen. Any premature disturbance to the hive can result in the colony rejecting her. The colony may even kill her, thinking the disturbance is somehow her fault. Play it safe and leave the hive alone for one week. During that time, worker bees eat through the candy and release the queen from her cage. She becomes the accepted leader of the colony.

As mentioned in Chapter 6, conduct your first inspection on a mild, sunny day (55 degrees F or more) with little or no wind. As always, visit your hive sometime between 10 a.m. and 5 p.m.

Smoke and open your hive, and remove the first frame. Place it vertically on the ground, leaning it up against the hive. Other than a few occasional bees, not much will be happening on this frame. In all likelihood, the bees haven't had time to draw the foundation into honeycomb.

As you continue your inspection of each subsequent frame, you should begin to see more and more going on. Toward the center of the hive you should see that the girls have been busily drawing out the wax foundation into honeycomb.

Verifying that the queen was released

When you reach the two frames sandwiching the queen cage, look down in the hole where the candy plug was. If the candy is gone, that's wonderful! It means worker bees have opened the cage and released the queen. Remove the cage and peek inside. Confirm that the queen has been released. Place the cage near the entrance so that any worker bees that you find exploring in the cage find their way back into the hive.

Removing any burr comb

You're likely to find that industrious bees have built lots of *burr comb* (sometimes called *natural comb, wild comb,* or *brace comb*) in the gap created by the queen cage. You may find comb on and around the queen cage itself.

Although it's beautiful bit of engineering, you must remove this bright white comb of perfectly symmetrical cells. Failing to do so is sure to create all kinds of headaches for you later in the season.

Burr comb refers to bits of random wax comb that connect two frames together or connect any hive parts together. Such comb is an extension of comb beyond what the bees build within the frames. Burr comb needs to be removed by the beekeeper to facilitate manipulation and inspection of frames.

Using your hive tool to sever the burr comb where it's connected to the frames, slowly lift the comb straight up and out of the hive. You'll probably find it covered with worker bees.

Examine the comb to ensure the queen isn't on it. If she is, you must gently remove her from the comb and place her back into the hive. Queens are quite easy to handle, and although they have a stinger, they're not inclined to use it. Simply wet your thumb and index finger and gently grasp her by her wings. She moves quickly, so it may take a few tries. Don't jab at her, but rather treat her as if she were made of eggshells. Easy does it!

Removing the bees from this burr comb is essential. One or two good shakes dislodge the bees from the comb. Shaking bees loose is a technique that will come in handy many times in the future.

Shaking is a sharp downward motion with an abrupt halt just above the frames.

Save the natural comb to study at your leisure back at home. Look for eggs, because the queen often starts laying on this comb. It makes a great "show and tell" for children! And you can always use beeswax to make cool things, like candles, furniture polish, and cosmetics (see Chapter 14 for some recipes).

Looking for eggs

Taking a close look at the frames that were near the queen cage, what do you see? Pollen? Nectar? Great! Do you see any eggs? They're the primary things you're looking for during your first inspection (see "Basic Inspection Techniques" earlier in this chapter). When eggs are present, you know the queen already is at work. That's all you need to find out on this first inspection. Close things up, and leave the bees alone for another week. Be satisfied that all is well. Because the weather likely is still cool, you don't want to expose the new colony to the elements for too long.

If no queen or eggs can be found, you may have a problem. In this abnormal situation, wait another few days and check once again. Seeing eggs is evidence enough that you have a queen, but if you still find no evidence of the queen, you need to order a new one from your bee supplier. The colony will do okay while awaiting its new queen, which you'll introduce exactly as you did the original: by hanging the cage between two frames and leaving the bees alone for a full week.

Replacing the tenth frame

The tenth frame is the one that you removed when you originally hived your package. It now becomes your *wall frame*.

Providing more syrup

When necessary, replenish your hive-top feeder with more sugar syrup. The recipe for sugar syrup can be found in Chapter 5.

The second and third weeks

On that first visit, you were looking for evidence that the queen had been released and was laying eggs. During the inspections that you conduct two and three weeks after hiving your package, you're trying to determine how well the queen is performing. By now there are a lot of new things to see and admire.

Following standard procedure, smoke, open the hive, and remove frames one by one for inspection. Work your way toward the center of the hive. As always, look for eggs. They're your ongoing assurance that the queen is in residence.

Note that the bees have drawn more of the foundation into honeycomb. They work from the center outward, so that the outer five to six frames haven't likely been drawn out yet. That's normal.

Looking for larvae

By the second week you can easily see larvae in various stages of development (see Figure 7-5). They should be bright white and glistening like snowy white shrimp! Looking closely, you may even witness a larva moving in its cell or spot a worker bee feeding one.

Evaluating your queen

Estimate how many eggs her majesty is laying. One good way to tell is if you have one or two frames with both sides ¾-filled with eggs and larvae. That means your queen is doing a super fantastic job. Congratulations!

If you have one or two frames with only one side filled, she's doing moderately well. If you find fewer than that, she's doing poorly, and you need to consider replacing her as soon as possible. See Chapter 9 for instructions on how to replace your queen.

Hunting for capped brood

By the third week you'll begin seeing capped brood — the final stage of the bees' metamorphosis. *Capped brood* are light tan in color, but note that the brood cappings on older comb are a darker tan. The capped brood are located on frames that are closest to the center of the hive. Cells with eggs and larvae are on the adjacent frames.

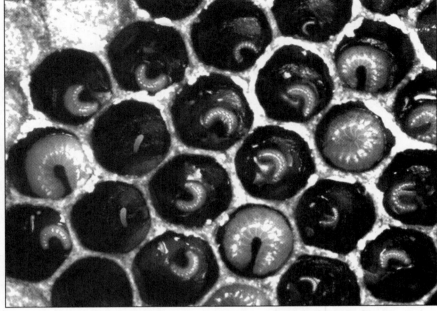

Figure 7-5:
Larvae go through various stages of development. Also, note eggs in this photo.

Courtesy of Dr. Edward Ross, California Academy of Sciences

An excellent queen lays eggs in nearly every cell, skipping few cells along the way and resulting in a pattern of eggs, larvae, and capped brood that is tightly packed together, stretching all the way across most of the frame.

You'll also notice a crescent of pollen above each capped brood, and a crescent of nectar or capped honey above the pollen. This is a picture perfect situation.

A spotty and loose brood pattern also can be evidence of a problem. You may have a poor queen, in which case she should be replaced as soon as possible. Sunken or perforated brood cappings may be evidence of disease, in which case you must diagnose the cause and take steps to medicate. (See Chapter 10 for more about bee diseases and remedies.)

Looking for supersedure cells

The third week also is when you need to start looking for *supersedure* cells (also called queen cells). The bees create supersedure cells if they believe their queen is not performing up-to-par. These peanut-shaped appendages are an indication that the colony may be planning to replace (or supersede) the queen. Queen cells located on the upper two-thirds of the frame are supersedure cells (see Figure 7-6). On the other hand, queen cells located on the lower third of the frame are not supersedure cells but are called swarm cells, which are discussed later in this chapter. **Note:** Swarming seldom is a problem with a new hive this early in the season.

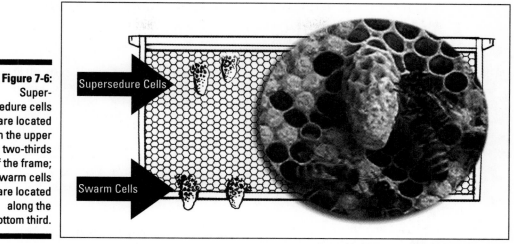

Figure 7-6: Super- sedure cells are located in the upper two-thirds of the frame; swarm cells are located along the bottom third.

Photo courtesy of David Eyre, beeworks.com

The bees create *swarm cells* to raise a new queen in preparation for the act of *swarming.* This usually happens when conditions in the hive become too crowded. The colony decides to split in half — with half the population leaving the hive (swarming) with the old queen and the remaining half staying behind with the makings for a new queen (the ones that are developing in the queen cells, or swarm cells).

If you spot more than three to four supersedure cells, you need to order a new queen, because giving the bees a new queen is far better than letting them create their own. Furthermore, you'll lose less time and guarantee a desirable lineage.

Supersedure is a natural occurrence when a colony replaces an old or ailing queen with a new queen.

Provide more syrup

Check every week to see whether your hive-top feeder has enough sugar syrup. Replenish as needed. The recipe for sugar syrup is in Chapter 5. Just pour it into the feeder.

Weeks four through eight

Things really are buzzing now that a month has passed since you hived your bees. Or at least they should be.

Perform your inspection as always, looking for evidence of the queen (eggs) and a good pattern of capped brood, pollen, and capped honey.

Adding a second deep hive body

If all's well, by the end of the fourth week the bees have drawn nearly all the foundation into comb. They've added wax produced in their wax glands to the foundation, creating the comb cells in which they store pollen, honey, and brood. When seven of the ten frames are drawn into the comb, you want to add your second deep hive body (see Chapter 4 for more information). Anticipate the need for this addition because timing is important. If you wait too long the colony may grow too fast (with up to 2000 new bees emerging every day!), become overcrowded, and eventually swarm. Add the second deep hive body too early and the colony below loses heat and the brood may become chilled and die.

When adding the second deep hive body becomes necessary, follow these steps:

1. **Smoke your hive as usual.**

2. **Remove the outer cover and the hive-top feeder (or the inner cover, if one is being used)**

3. **Place the second deep directly on top of the original hive body.**

4. **Fill the new second story with ten frames and foundation (see Chapter 4).**

5. **Put the hive-top feeder directly on top of the new upper deep, and below the outer cover.**

   Replenish sugar syrup if needed.

6. **Replace the outer cover.**

The upper deep will be used during the early summer for raising brood. But later on it serves as the food chamber for storing honey and pollen for the upcoming winter season.

Witnessing a miracle!

By the fifth week the frames are jam-packed with eggs, larvae, capped brood, pollen, and honey. Look carefully at the capped brood. You may see a miracle in the making. Watch for movement under the capping. A new bee is about to emerge! She'll chew her way out of the cell and crawl out (see Figure 7-7). At first, she totters about, looking like a newborn, yet she quickly learns how to use her legs. She appears lighter in color than her sisters and is covered with soft, damp hairs. Her eyes are tiny. What a joy this is to witness. Savor the moment!

Courtesy of John Clayton

Figure 7-7:
This young adult bee is just emerging from her cell.

Watching for swarm cells

During weeks six through eight, continue looking for queen cells, but you also want to be on the lookout for cells in the lower third of the frames. They are swarm cells (as mentioned previously in this chapter), which are an early indication that the hive may consider swarming. During your first season, don't be too concerned if you spot an occasional swarm cell. It isn't likely that a new colony will swarm. However, when you find eight or more of these swarm cells, you can be fairly certain the colony intends to swarm.

You don't want a swarm to happen. When a colony swarms, half the population leaves with the old queen looking for a new and more spacious home. Before that happens, the bees take steps — evidenced by the presence of swarm cells — to create a new queen. But with half of the girls gone, and several weeks lost while the new virgin queen gets up to speed, you're left with far fewer bees gathering honey for you. Your harvest will be only a fraction of what it might otherwise have been. Prevent this unhappy situation before it happens by anticipating the bees' need for more space and adequate ventilation. (See Chapter 9 for more about swarming.)

Providing more ventilation

During weeks five or six, you need to improve hive ventilation by opening the hive entrance. Turn the entrance reducer so that the larger of its two openings is in position. The larger opening is about 4 inches wide. (Eventually, the reducer is completely removed.) The colony is now robust enough to protect itself, and the weather is milder.

You can remove the entrance reducer completely in the eighth week following the installation of your bees.

What to do about propolis during inspections

Here's a question I'm often asked: "My bees have built comb on and around some of the frames. There's comb along the bottom bars, and they've glued some of the frames together with propolis. Should I scrape this wax and propolis off, or should I leave things as the bees intended?"

With every inspection you should take a moment to scrape extra wax and propolis off the frames. Don't let it build up or the job will become too daunting to do anything about it. Take a few minutes during every inspection to tidy things up. Getting into the habit of cleaning house will save you tons of work later on. If you allow the bees to glue everything together, it's tempting to forgo a necessary inspection to avoid the challenge of pulling things apart. Don't let the bees get the upper hand. Scrape it off!

Manipulating the frames of foundation

By the seventh or eighth week, manipulate the order of frames to encourage the bees to draw out more foundation into comb cells. You can do this by placing any frames of foundation that haven't been drawn between frames of newly drawn comb. However, don't place these frames smack in the middle of the brood nest. That would be counterproductive because doing so splits (or breaks) the nest apart, making it difficult for bees to regulate the environment of the temperature-sensitive brood.

Making room for honey!

As the eighth week approaches, you may find that the bees may have drawn out seven of the ten frames in the upper deep. When that happens, remove the hive-top feeder and add a queen excluder and a shallow honey super with frames and foundation (see Chapter 4 for more information on woodenware). The girls now are ready to start collecting honey for you!

The act of adding shallow (honey) supers to a colony is called *supering*.

Chapter 8

Different Seasons, Different Activities

. . . .

In This Chapter

▶ Beekeeping chores for summer, autumn, winter, and spring

▶ Starting your second season

. . . .

*T*he seasonal calendar of events in Maine obviously looks different than one in southern California. But different climates mean different schedules and activities for the hive and beekeeper. Regardless of their precise location, honey bees are impacted by the general change of seasons. Knowing what major activities are taking place within the hive and what's expected of you during these seasons is useful. For a good beekeeper, *anticipation* is the key to success.

This chapter contains a suggested schedule of seasonal activities for the beekeeper. However, you must note that geography, weather, climate, neighborhood, and even the type of bees influence the timing of these activities. Nevertheless, this chapter gives you a generic overview of what's going on in the hive during each season.

I also suggest some important tasks for the beekeeper and provide a rough estimate of the amount of time that you'll need to spend with your bees during each season. These time estimates are based on maintaining one or two hives.

Buzzing through those Lazy, Hazy, Crazy Days of Summer

Nectar flow usually reaches its peak during summer. That's also when the population of the colony usually reaches its peak. When that's the case, your colonies are quite self sufficient, boiling with worker bees tirelessly collecting pollen, gathering nectar, and making honey. Note, however, that the queen's rate of egg laying drops a bit during the late summer.

On hot and humid nights, you may see a huge curtain of bees hanging on the exterior of the hive. Don't worry. They're simply cooling off on the front porch.

Late in summer the colony's growth begins to diminish. Drones still are around, but outside activity begins slowing down when the nectar flow slows. Bees seem to be restless and become protective of their honey.

Your summer "to-do" list

Here are some activities you can expect to schedule between trips to the beach and hot-dog picnics.

- ✔ Inspecting the hive every other week, making sure that it's healthy and that the queen is present.

- ✔ Adding honey supers as needed. Keep your fingers crossed in anticipation of a great honey harvest.

- ✔ Keeping up swarm control through mid-summer (see Chapters 7 and 9). Late in the summer there is little chance of swarming.

- ✔ Being on the lookout for honey-robbing wasps or other bees. A hive under full attack is a nasty situation (see Chapter 9 for information about how to deal with robbing).

- ✔ Harvesting your honey crop at the end of the nectar flow (see Chapters 12 and 13). Remember that the colony requires at least 60 pounds of honey for use during winter. This is the time to break out your gloves, because your normally docile bees are at their most defensive. They don't want to give up their honey without a bit of a fight!

Your summer time commitment

You can't do all that much until the end of the summer and the honey harvest, because your bees are doing it all! Figure on spending about six to eight hours with your bees during the summer months. Most of this time involves harvesting and bottling honey (see Chapters 12 and 13 for more information on honey harvesting).

Falling Leaves Point to Autumn Chores

Most nectar and pollen sources become scarce as days become shorter and weather cools in autumn. All in all, as the season slows down, so do the activities within your hive: The queen's egg laying is dramatically reduced, drones begin to disappear from the hive, and hive population drops significantly.

Making your winter ventilation preparations a breeze

Here's an easy ventilation trick from a commercial beekeeper who has successfully overwintered thousands of hives in upper New York State. During your late autumn preparation, simply slide the upper deep back so that you create a ⅛-inch opening along the entire front of the lower hive body. Don't make it a larger gap, or the bees will use it as an entrance or you might create a robbing situation. "Wait a minute!" you might say. "Doesn't the rain get in that little gap?" Yes it does. But that's no problem because you've already tilted your hive slightly forward (see "Picking the Best Location" in Chapter 3). Any rain or snow that dribbles in simply drains right out the front door. Try this trick along with your other ventilation routines.

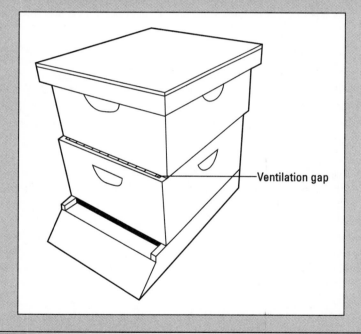

Ventilation gap

Your bees begin bringing in propolis, using it to chink up cracks in the hive that may leak the winter's cold wind. The colony is hunkering down for the winter, so you must help your bees get ready.

Watch out for robbing during this time (other bees would love to steal honey from your hives). For more about robbing and how to prevent it, see Chapter 9.

Your autumn "to-do" list

When helping your bees prepare for the upcoming hardships of winter months, you must

✔ Inspect your bees (look inside the hive) and make certain that the queen is there. As mentioned in Chapter 7, the easiest way is finding eggs. One egg per cell means the queen is present.

Be sure to look for eggs, *not* larvae. Finding eggs means that the queen was present two days ago. Larvae, on the other hand, can be three to eight days old. Thus, merely finding larvae is no guarantee that you have a queen.

When you wait too late during autumn, you discover that eggs and larvae are few and far between. In that case, actually finding the queen is the surest way to check. Be patient, and look carefully.

✔ Determine whether the bees have enough honey. Your bees need plenty of food (capped honey) for the winter. Make certain that the upper deep-hive body is full of honey. Honey is essential for your bees' survival, because it's the fuel that stokes their stoves. Without it they're certain to perish.

In cooler, northern climates, hives need about 60 to 70 pounds of honey headed into winter. Less honey reserves (30 to 40 pounds) if your winters are short (or nonexistent).

✔ Feed and medicate your colony. They'll accept a 2-to-1 sugar-syrup feeding (see "Winter syrup recipe" sidebar) until colder weather contracts them into a tight cluster. At that point, temperatures are too cold for them to leave the cluster ("see the Winter Wonderland" section later in this chapter), so feeding them is useless.

Keep feeding your bees until they stop taking the syrup, or until the temperature drops and they form the winter cluster. A hive-top feeder works best. The first gallon should be medicated with Fumidil — subsequent feedings are not medicated.

✔ Provide adequate ventilation. During winter, the temperature at the center of the cluster is maintained at 90 to 93 degrees F. Without adequate ventilation, the warm air from the cluster rises, hits the cold inner cover, and condensation drips down onto the bees as ice-cold water. That's a big problem! The bees will become chilled and die. Keep your colony dry by doing the following:

• Gluing (permanently) four postage stamp-sized pieces of wood (you can use the thin end of a wood shingle or pieces of a Popsicle stick) to the four corners of the inner cover's flat underside. This neat ventilation trick makes an air space of $\frac{1}{16}$ inch or less between the top edge of the upper deep-hive body and the inner cover.

- Placing the inner cover on the top deep body, *flat side down.* The oval hole should be left open, and the notch in the ledge of the inner cover should be left open for ventilation.

When you put the outer cover on the hive, make sure that you push it forward so the opening in the inner cover remains open. Make sure that the outer cover is put on the hive equidistant from side to side. The result is a gentle flow of air that carries off moisture from the underside of the inner cover and thus keeps the colony dry.

✔ Wrap the hive in black tar paper (the kind used by roofers, see Figure 8-1) if you're in a climate where the winter gets below freezing for more than several weeks. Make sure that you don't cover the entrance or any upper ventilation holes. The black tar paper absorbs heat from the winter sun, and helps the colony better regulate temperatures during cold spells. It also acts as a windbreak.

I put a double thickness of tar paper over the top of the hive. Placing a rock on top ensures that cold winds don't lift the tar paper off. I also cut a hole in the wrapping to accommodate the ventilation hole I drilled in the upper deep-hive body (see Figure 8-1).

✔ Provide a windbreak if your winter weather is harsh. It is hoped that you originally were able to locate your hives with a natural windbreak of shrubbery (see Chapter 3). But if not, you can erect a temporary windbreak of fence posts and burlap. Position it to block prevailing winter winds.

Winter syrup recipe

I use special syrup for feeding bees that are going into the winter months. The thicker consistency recipe makes it easier for bees to convert the syrup into the honey they'll store for the winter.

Boil 2½ quarts of water on the stove. When it comes to a rolling boil, turn off the heat and add 10 pounds of white granulated sugar. Be sure that you turn off the stove. If you continue boiling the sugar, it may caramelize, and that makes the bees sick. Stir until the sugar dissolves completely. The syrup must cool to room temperature before you can add medication. Note

that this is a recipe for thicker syrup than that used for feeding bees in the spring.

For medicated syrup, mix 1 teaspoon of Fumidil-B in approximately a half a cup of cool water (the medication won't dissolve in the syrup). Fumidil-B protects your bees against nosema — a common bee illness (see Chapter 10 for more information on nosema and other diseases). Add the medication to the syrup and stir. You also can add two tablespoons of Honey B Healthy. This food supplement contains essential oils and has a number of beneficial qualities.

Your autumn time commitment

Figure on spending three to five hours total to get your bees fed, medicated, and bedded down for the winter months ahead.

Clustering in a Winter Wonderland

What goes on in a beehive during winter? The queen is surrounded by thousand of her workers — kept warm in the midst of the winter cluster. The winter cluster starts in the brood chamber when ambient temperatures reach 54 to 57 degrees F. When cold weather comes, the cluster forms in the center of the two hive bodies. It covers the *top* bars of the frames in the lower chamber and extends over and beyond the *bottom* bars of the frames in the food chamber (see Figure 8-2).

Although the temperature outside may be freezing, the center of the winter cluster remains a constant 92 degrees F. The bees generate heat by "shivering" their wing muscles.

No drones are in the hive during winter, but some worker brood begin appearing late in the winter. Meanwhile, the bees consume about 50 to 60 pounds of honey in the hive during winter months. They eat while they are in the cluster, moving around as a cluster whenever the temperature gets above 40 to 45 degrees F. They can move to a new area of honey only when the weather is warm enough for them to break cluster.

Figure 8-2:
Although the outside temperature may be freezing, the center of the winter cluster remains a balmy 92 degrees F. This cutaway illustration shows the the winter cluster's position.

Bees won't defecate in the hive. Instead they hold off until they can leave the hive on a nice, mild day when the temperature is 45 to 50 degrees F to take *cleansing flights*.

Your winter "to-do" list

Winter is the slowest season of your beekeeping cycle. You've already prepared your colony for the kinds of weather that your part of the world typically experiences. So, now is the time to do the following:

- Monitor the hive entrance. Brush off any dead bees or snow that block the entrance.

- Make sure the bees have enough food! The late winter and early spring is when colonies can die of starvation.

    Late in the winter, on a nice, mild day when there is no wind and bees are flying, take a quick peek inside your hive. It's best not to remove any frames. Just have a look-see under the cover. Do you see bees? They still should be in a cluster in the upper deep. Are they okay?

    If you don't notice any sealed honey in the top frames, you may need to begin some emergency feeding. But remember that once you start feeding, you *cannot* stop until the bees are bringing in their own pollen and nectar.

✔ Clean, repair, and store your equipment for the winter.

✔ Attend bee club meetings, and read all those back issues of your favorite bee journals.

✔ Order package bees and equipment (if needed) from a reputable supplier.

✔ Try a bee-related hobby. The winter is a good time for making beeswax candles, brewing some mead, and dreaming of spring! See Chapter 14 for some ideas.

Your winter time commitment

Not much is doing with bees during winter. They are in their winter cluster, toasty and warm inside the hive. Figure on spending two to three hours repairing stored equipment, plus whatever time you may spend on bee-related hobbies — making candles, mead, cosmetics, and so on — or attending bee-club meetings. You might even decorate your hive for the holidays (see Figure 8-3). Just don't cover the ventilation holes!

Figure 8-3:
In the season to be jolly, deck your hive with boughs of holly!

Spring is in the Air (Starting Your Second Season)

Spring is one of the busiest times of year for bees (and beekeepers). It's the season when new colonies are started and established colonies come "back to life."

Days are getting longer and milder, and the established hive comes alive, exploding in population. The queen steadily lays more and more eggs, ultimately reaching her greatest rate of egg laying. The drones begin reappearing and hive activity starts hopping. The nectar and pollen begins coming into the hive thick and fast. The hive boils with activity.

Your spring "to-do" list

Beekeepers face many chores in the springtime, evaluating the status of their colonies and helping their bees get into shape for summer months. Some of those chores include

- Conducting an early bird inspection. Colonies should be given a quick inspection as early in the spring as possible. The exact timing depends upon your location (earlier in warmer zones, later in colder zones).

  You don't need to wait until bees are flying freely every day nor until the signs of spring are visible (the appearance of buds and flowers). Do your first spring inspection on a sunny, mild day with *no wind* and a temperature close to 50 degrees F.

  A rule of thumb: If the weather is cold enough that you need a heavy overcoat, it's too cold to inspect the bees.

- Determining whether your bees made it through the winter. Do you see the cluster? The clustered bees should be fairly high in the upper deep-hive body. If you don't see them, can you hear the cluster? Tapping the side of the hive and putting your ear against it, listen for a hum or buzzing.

  If it appears that you've lost your bees, take the hive apart and clean out any dead bees. Reassemble it and order a package of bees as soon as possible. Don't give up. We all lose our bees at one time or another.

- Checking to make sure that you have a queen. Look down between some of the frames. (Do you see any brood?) That's a good sign that the queen is present. To get a better look, you may need to carefully remove a frame from the center of the top deep. Can you see any brood? Do you spot any eggs?

This inspection must be done quickly, because you don't want to leave the frame open to chilly air. If you don't see any brood or eggs, your hive may be without a queen, and you should order a new queen as soon as possible, assuming, that is, the hive population is sufficient to incubate brood once the new queen arrives. What's sufficient? The cluster of bees needs to be *at least* the size of a large grapefruit (hopefully larger). If you have fewer bees than that, you should plan to order a new package of bees (with queen).

✔ Checking to ensure the bees still have food. Looking down between the frames, see if you spot any honey. Honey is capped with white cappings (tan cappings are the brood). If you see honey, that's great. If not, you must begin emergency feeding your bees (see the following bulleted items).

✔ Medicating and feeding the colony. A few weeks before the first blossoms appear, you need to begin medicating and feeding your bees (regardless of whether they still have honey).

Feed the colony sugar syrup (see recipe in Chapter 5). This feeding stimulates the queen and encourages her to start laying eggs at a brisk rate. The first gallon needs to be medicated with Fumidil — subsequent gallons aren't medicated. See the section, "Administering spring medication," later in this chapter. Continue feeding until you notice that the bees are bringing in their own food. You'll know when you see pollen on their legs.

Feed the colony pollen substitute, which helps strengthen your hive and stimulates egg laying in the queen. Pollen substitute is available in a powdered mix from your bee supplier. This feeding can cease when you see bees bringing in their own pollen.

✔ Reversing your hive bodies (see "Reversing hive bodies" section later in this chapter).

✔ Anticipating colony growth. Don't wait until your hive is "boiling" with bees. Later in the spring, before the colony becomes too crowded, create more room for the bees by adding a queen excluder and honey supers. Be sure that you remove the feeder and discontinue all medication at this time.

✔ Watching out for indications of swarming. Inspect the hive periodically and look for *swarm* cells (see Chapter 7).

Your spring time commitment

Spring is just about the busiest time for the beekeeper. You can anticipate spending eight to twelve hours tending to your bees.

Administering spring medication

Although you probably don't need to medicate your bees during their first season, you'll definitely want to begin medication in the spring of your colony's second season. Remember to stop all medication treatments five to six weeks *before* adding honey supers to the colony to prevent contamination of the honey that you want to harvest.

The list that follows contains a springtime medication regime that helps prevent diseases, control mites, and improve your bees' overall health (see Chapters 9, 10, and 11 for more information):

- ✔ **To prevent nosema:** In a small jar half filled with lukewarm water, add one teaspoon of *Fumidil*. Shake the jar until dissolved. Stir the jar's contents into 1 gallon of the cooled sugar syrup solution you use to feed your bees (see Chapter 5). Feed at the top of the hive using a hive-top feeder.

- ✔ **To prevent foulbrood:** *Terramycin* is an antibiotic that comes as a powder. It's effective against American and European foulbrood. For each colony, mix 1 teaspoon of Terramycin with 2 tablespoons of *powdered* sugar (do not use granulated sugar). Sprinkle the sugar mixture on the *ends* of the frames' top bars. Don't sprinkle where it comes in direct contact with open brood cells, because it's toxic to larvae. Repeat this "dusting" two more times at three- to five-day intervals.

- ✔ **For general health:** *Honey B Healthy* contains essential oils (lemongrass and spearmint). The beneficial properties of using essential oils in hives are well documented. Use it as a feeding stimulant by adding a teaspoon of the concentrate to your sugar syrup solution during your spring feedings. It helps keep bees healthy even in the presence of mites.

- ✔ **For varroa mite control:** Place *Apistan* strips within the hive, using one strip for every five frames of bees. (See Chapter 11 for more information on using Apistan.) If two deep-hive bodies are used for the brood nest,

Starting and stopping sugar syrup feedings

Continue feeding the bees sugar syrup in the spring until they stop taking the syrup — or until it is evident that the bees are bringing in nectar. The exception is for a newly established colony — in which case you should continue feeding syrup until all of the frames of foundation are drawn into comb, or until the bees stop feeding on the syrup — whichever comes first. In the autumn, continue feeding until they stop taking the syrup, or the daytime temperature drops to less than 40° F (4° C) — whichever comes first.

hang Apistan strips in alternate corners of the cluster, utilizing the top and bottom hive bodies. Make certain each strip is placed between the frames, not on top of them. For best chemical distribution, use Apistan when daytime high temperatures are at least 50 degrees F. Keep strips in the hive for six to eight weeks. Don't remove strips from the hive for at least 42 days (six weeks). Don't leave strips in the hive for any more than 56 days (eight weeks). Honey supers may be put on the hive after the strips are removed.

Never ever leave Apistan in the hive over the winter. Doing so constantly exposes the mites to the active ingredient, which becomes weaker and weaker over time. These sublethal doses increase the chance for mites to build up a resistance to Apistan. This tolerance is then passed on to future generations of mites, and subsequent treatments become useless.

✔ **For tracheal mite control:** When the weather starts getting warmer, place a prepared bag containing 1.8 ounces of *menthol crystals* on the top bars toward the rear of the hive (see Chapter 11 for more information). Set them on a small piece of aluminum foil to prevent the bees from chewing holes in the bag and carrying it away. Leave the bag in the hive for 14 *consecutive* days when the outdoor temperature ranges between 60 and 80 degrees F.

Adding a grease patty to the top bars of the brood chamber is another treatment for tracheal mites. Making grease patties is easy; see Chapter 11 for the recipe that I use. Use one patty per hive, replacing them as the bees consume them. Use these patties throughout the year (even when honey supers are on the hive). Unused patties can be stored in the freezer until you're ready to use them.

Reversing hive bodies

Bees normally move upward in the hive during the winter. In early spring, the upper deep is full of bees, new brood, and food. But the lower deep-hive body is mostly empty. You can help matters by reversing the top and bottom deep-hive bodies (see Figure 8-4). Doing so also gives you an opportunity to clean the bottom board. Follow these steps:

1. **When a mild day comes along (50 degrees F) with little or no wind and bright clear sunlight, open your hive using your smoker in the usual way.**

2. **Place the upturned outer cover on the ground and then remove the upper deep-hive body.**

3. **Keep the inner cover on the deep and close the oval hole in the middle of the inner cover with a piece of wood shingle or tape.**

4. **Place the deep across the edges of the outer cover, so there will be only four points of contact (you'll squeeze fewer bees that way).**

5. **Now you can see down into the lower deep that still rests on the bottom board.**

It probably is empty, but even if some inhabitants are found, lift the lower deep off the bottom board and place it crossways on the inner cover that is covering the deep you previously removed.

6. **Scrape and clean the bottom board.**

Note: This is good opportunity to add a slatted rack (see Chapter 4), because you won't get another chance until autumn. Slatted racks help with the hive's ventilation and can promote superior brood patterns. They also encourage the queen to lay eggs all the way to the front of the hive, because of improved ventilation and draft control.

7. **Now stand the deep body — which had been the relatively empty bottom one — on one end, placing it on the ground.**

Then place the *full* hive body onto the clean bottom board (or on the slatted rack, if you added one).

8. **Smoke the bees and remove the inner cover so that you can place the empty deep on top.**

Replace the inner and outer covers.

This reversing procedure enables the bees to better distribute brood, honey, pollen, fresh nectar, and water. Reversing gives them more room to move upward, which is the direction that they always want to move.

Repeat this reversal in about three to four weeks, restoring the hive to its original configuration. At that time you can put on one or more honey supers — assuming the bees are now bringing in their own food and you have ceased feeding and medicating.

Figure 8-4:
Reversing hive bodies in the spring helps to better distribute brood and food, and speeds up the growth of your colony's population.

Part IV
Common Problems & Simple Solutions

The 5th Wave   By Rich Tennant

"I try to see to it that it's not all work for my bees."

In this part . . .

Sometimes things go wrong. But don't worry. In these chapters, I tell you what to expect and what to do when things don't go as planned. Find out how to keep your bees from swarming, getting sick, or undergoing stress. If you encounter a problem, refer to this section for the solution.

Chapter 9

Anticipating and Preventing Potential Problems

. . . .

In This Chapter

▶ Preventing and controlling swarming

▶ Replacing your queen

▶ Thwarting robbing frenzies

▶ Ridding your hive of laying workers

▶ Preventing pesticide poisoning

▶ Understanding the "killer bee" phenomenon

. . . .

Despite the best intentions and the most careful planning, things occasionally go wrong. It happens. The bees swarm. The queen is nowhere to be found. The whole colony dies or flies away. What happened? What did you do wrong? What should you have done differently?

I've made just about every mistake in the book at one time or another. But that's nothing to be ashamed of. It's part of the learning process. The key lesson I've learned has been to *anticipate*. Discipline yourself to plan ahead and look out for potential problems *before* they happen. I can assure you that you can head off 80 to 90 percent of potential problems at the pass if you anticipate trouble and take steps to avoid it.

In this chapter, I include a few of the more common nonmedical problems to anticipate and avoid. These problems include swarming and absconding, losing your queen, and losing your colony because of cold, robbers (robber bees, that is!), and pesticides. This chapter also tells you how to deal with potential community-mindset problems of having Africanized bees in your geographical area.

Running Away (To Join the Circus?)

Sometimes bees disappear. They simply get up and go. Poof! In one scenario, called *swarming*, about 50 percent of the colony packs up with the queen and takes flight. In the other scenario, called *absconding*, 100 percent of the colony hits the road, leaving not a soul behind. Neither scenario is something you want to happen.

Swarming

A swarm of honey bees is a familiar sight in the spring and early summer. It's one of the most fascinating phenomena in nature and an instinctive way that honey bees manage the colony's growth and survival. To witness a swarm pouring out of a hive is simply thrilling — though the pleasure may be less so if the swarm of bees is yours!

Immediately before swarming, the bees that intend to leave the colony gorge themselves with honey (like packing a box lunch before a long trip). Then, all at once, like someone flipped a switch, tens of thousands of bees exit the hive and blacken the sky with their numbers. Half or more of the colony leaves the hive to look for a new home. But first, within a few minutes of departing from the hive, the bees settle down on a nearby surface.

There's no telling where a swarm might land. It could land on any convenient resting place: a bush, a tree branch (see Figure 9-1), a lamppost, or perhaps a piece of patio furniture (see Figure 9-2). In any case, the swarming bees won't stay there long. As soon as scout bees find a more suitable and protected home, the swarm will be up, up, and away.

What's the buzz?

You may have a friend who knew a couple who kept hearing a mysterious humming noise in their bedroom at night. After a while, honey started oozing out from under the wall. Upon taking the wall apart, they discovered the whole space within the wall had been turned into a hive. No, this is not an urban legend! I get calls like this every year — at some point a colony of bees has swarmed and set up housekeeping in a cozy niche of a human habitation. And now there's honey dripping through the ceiling and so forth. This is one whacking big job to deal with and very expensive. The walls have to be torn apart to get at such hives for removal.

Courtesy of John Clayton

In its temporary resting place, the swarm is a bundle of bees clustered together for protection and warmth. In the center of it all is their queen. Depending on the size of the hive that swarmed, the cluster may be as small as a grapefruit or as large as a watermelon. The bees will remain in this manner for a few hours or even a few days while scout bees look for a new home. When they return with news of a suitable spot, off they all go to take up residence in a hollow tree, within the walls of an old barn, or in some other cozy cavity.

Not sure if your hive has swarmed? A regular inspection during the month of May will reveal the situation. Know the key indicators: no eggs, fewer bees, and all the cells have only older larvae and/or capped brood.

Understanding why you want to prevent swarming

Swarms are a dramatic sight, to be sure, but they are not good news for you. A colony that swarms is far less likely to collect a surplus of honey. That means no honey harvest for you. A colony that loses 50 percent of its population and 50 percent of its honey also will have a difficult time regaining its population and productivity. That means the bees will have a tougher time making it through the winter months.

Figure 9-2:
A swarm
that has
taken up
temporary
residence
under a
picnic table.

Courtesy of John Clayton

It's bad enough news when your bees swarm, but the *later* in the season they do it, the worse the news is for you. If the bees choose to swarm later, there simply isn't enough time for the colony to recover during that season.

There's an old poem of unknown origin that is well known to beekeepers:

> A swarm in May — is worth a load of hay.
>
> A swarm in June — is worth a silver spoon.
>
> A swarm in July — isn't worth a fly.

If you're a first-year beekeeper, rest assured that a new colony is unlikely to swarm during its first season. But older and more crowded colonies are likely candidates for swarming behavior. Remember, swarming is a natural and normal instinct for bees. At one point or another, your bees will want to swarm. Preventing them from doing so is a skill every beekeeper should learn.

Keeping the girls from leaving home

There are two primary reasons bees swarm: congestion and poor ventilation. Occasionally, a poorly performing queen can contribute to the swarming impulse. But all these conditions can be anticipated and avoided. Here are some things you can do:

✔ **Avoid congestion.** Because overcrowding is a primary reason a colony will swarm, make sure to anticipate your bees' needs and provide them with more room *before* they need it. If you wait until it's obvious that the colony is crowded, you're too late! The colony is likely to swarm. You can do the following to prevent congestion:

- Reverse your hive bodies in the early spring to better distribute the fast-growing population (see Chapter 8).

- Add a queen excluder and honey supers *before* the first nectar flow in the early spring (stop feeding and medicating before you add honey supers; see Chapter 8).

✔ **Provide adequate ventilation.** To ensure proper ventilation, you can do a number of things:

- Make sure that the notched ventilation hole in the front of the inner cover is open. Stand at the rear of the hive and push the outer cover forward. Doing so prevents the overhang of the outer cover from blocking the hole.

- Glue a short length of a wooden Popsicle stick to each of the four corners of the inner cover. By doing so, you create a thin gap between the inner cover and the hive and improve airflow into and out of the hive. (Alternatively, you can place a short screw with a fat, domed head in each corner. The fat head of the screw creates the gap you want.)

- Drill wine cork–sized holes in your upper deep (below the hand hold) and in all your honey supers, as shown in Figure 9-3. Doing so not only provides extra ventilation but also provides the bees with additional entrances. You can control airflow by blocking and opening these holes as needed with corks or strips of tape. Be sure to close off these entrances in the wintertime and in the case of a new colony whose population is still too small to defend all these extra openings.

✔ **Make the bees comfortable in hot weather by doing the following:**

- Supply a nearby water source. The bees will use this water to regulate the hive's temperature. See Chapter 3 for suggestions regarding water sources.

- Shield the hive from a full day of blazing sun. Locating the hive in dappled sunlight is the best solution (see Chapter 3).

✔ **Remove queen swarm cells — all of them.** The earliest evidence that your bees are thinking about swarming is that they start to make swarm cells (see Chapter 7). During the spring and early summer, inspect your hive every week or ten days to look for swarm cells. If you see any, remove them by cutting them out with the sharp end of your hive tool. The colony won't swarm if it doesn't have a new queen in the making.

WARNING!

This technique only works if you remove 100 percent of the swarm cells. If just one cell remains behind, the colony has the green light to swarm.

✔ **Replace your queen every other autumn.** Colonies with young queens are far less likely to swarm.

Figure 9-3: A useful way to provide a colony with ventilation is to drill wine cork–sized holes in the hive bodies and supers.

TIP

If the hive is simply boiling over with bees and you failed to take any of the above precautions, there is a last-resort emergency measure. You can remove all the frames of capped brood from the hive (with bees still on the frames) and replace them with frames of foundation. *A colony will not swarm if it has no capped brood.* Here's where having an empty nuc box is handy. The nuc gives you a place to house the frames, brood, and bees you remove.

The 7/10 rule

If you're a first-year beekeeper, here's a way to remember when it's time to give a new colony more room (and do so *before* it's too late):

✔ When 7 of the 10 frames in the lower deep are drawn into comb, add a second deep-hive body with frames and foundation.

✔ When 7 of the 10 frames in the upper deep are drawn into comb, add a queen excluder and a honey super.

✔ When 7 of the 10 frames in the honey super are drawn into comb, add an additional honey super.

Continue providing more room in this manner, adding more space when the bees have drawn out 70 percent of the foundation.

Make sure that the queen is not on any of these frames. You can use these frames of bees and brood to start a *new* hive! If there are eggs on those frames, the "new" hive will raise a new queen. Or you can play it safe and order a new queen from your bee supplier.

They swarmed anyway. Now what?

Okay, somehow you blew it. The bees swarmed anyway. You're not alone; it has happened to the best of us. The good news is that you may be able to capture your swarm and start another colony. (See the following section titled "Capturing a swarm.") You wanted a new hive of bees anyway, didn't you?

In any event, what should you do with the half of the colony that remains? Follow these steps:

1. **A week after your colony swarms, inspect the hive to determine whether you have a new queen.**

   You might spot a queen cell or two along the lower third of the frames (see Chapter 7 for tips on finding queen cells). That's an encouraging first step. It means a new queen is "in the oven." But you must ultimately determine if the colony's new queen is laying eggs. One week after a swarm you're unlikely to see any eggs — it's too soon for the new queen to get to work. But do have a look and see if you can find her majesty. If you can, great! Close up the hive and wait another week. If you *don't* see the queen, wait a couple more days and have another look.

   After the swarm, it will take six to eight days for the queen cell to open and a new virgin queen to emerge. Then allow three to four more days for her to mate with the drones. After another three to four days, she will start laying eggs. The total elapsed time since the swarm is about two weeks.

2. **Two weeks after the swarm, open the hive again and look for eggs.**

   Do you see eggs? If so, you have a queen and your colony is off and running. Close things up and celebrate with a glass of mead. If there's still no sign of a queen or her eggs, order a new queen from your bee supplier. Hive the replacement queen as soon as she arrives (see "Introducing a new queen to the hive" later in this chapter).

If you don't follow up after a swarm, the colony can easily become queenless without you ever being aware of it. No queen, no brood. No brood, no good.

Capturing a swarm

If your bees do swarm and you can see where they landed (and you can reach it safely), you can capture them and start a new hive. You may even be lucky enough to get a call from a friend or neighbor who has spotted a wild swarm in his yard (beekeepers are often called to come capture swarms). Either way, capturing a swarm is a thrilling experience.

Using an artificial swarm to prevent a natural swarm

There's another way to prevent a crowded hive from swarming: by creating an "artificial swarm" (sometimes called a "shook swarm"). This little trick is a bit of work, but it's effective because it gets the urge to swarm out of your colony's system. If you've tried everything else and you still think that the colony may swarm, here's how you can foil your bees' natural instincts:

1. Set up a new, empty hive about 10 feet from the hive you suspect is about to swarm. (I'll refer to this as the "new" hive.)

2. Put nine new frames with foundation into the new hive.

3. Turn your attention to the suspect hive. (I'll refer to this as the "old" hive.) Smoke and inspect the old hive, looking for the frame with the queen on it. When you find that precious frame, gently put it aside. Be careful! The queen is on that frame! If you have a spare hive body or an empty nuc, place this frame in it temporarily. You can even make use of a frame perch to hold this frame out of harm's way. In any event, find a way to keep the queen safe and sound while you tend to other things.

4. Place a bed sheet in front of the old hive, from the ground to the entrance board. You are creating a ramp for the bees that you are about to unceremoniously dump in front of this hive.

5. Take each frame out of the old hive, one by one, and shake off 80 to 90 percent of the bees. Shake them directly in front of the entrance of the old hive. The bed sheet will help them march their way back into the old hive (see the following figure).

6. Put these frames (with some bees still clinging to them) into the new, empty hive one by one. At this point, the new hive has nine frames containing brood, larvae, eggs, and about 10 to 20 percent of the bees from the old hive.

7. Add a tenth frame with new foundation to the new hive. Place it at either outside wall of the hive — it doesn't matter which.

8. Now back to the old hive. Take that precious frame with the queen on it that you set aside and shake all the bees (including the queen) off the frame in front of the old hive. The bees will walk back into their old home with their queen.

9. Put this frame into the old hive. Your old hive now contains this frame, nine new frames with foundation, about 80 to 90 percent of the bees, and their queen.

Your new hive now contains nine frames of brood, larvae, eggs, and about 10 to 20 percent of the bees from the old hive. The bees will use some of these eggs to raise a new queen (or you can speed things along and purchase a new queen from your bee supplier). The brood in this new hive will hatch soon, and this colony will be thriving in no time at all.

"Wait a minute!" you say. "I don't want another hive!" That's okay. After two weeks, you can recombine the new colony with the old. You'll have one big happy family again, and the bees will have lost their urge to swarm (you've taken care of that for them). To recombine the two colonies, the newspaper method is easiest (see Chapter 14).

Despite their rather awesome appearance, swarms are not dangerous. That's because honey bees are defensive only in the vicinity of their nest. They need this defensive behavior to protect their brood and food supply. But a swarm of honey bees has neither young nor food and is usually very gentle. (See this book's color section, "Things to Look For," to see how easily a swarm can be handled.) That's good news because it makes your job easy if you want to capture a swarm of bees.<Warning>

If you live in an area known to have Africanized Honey Bees (discussed later in this chapter), you must be very cautious — a swarm might be this undesirable strain. There's no way of telling just by looking at them. If you're in doubt, don't attempt to capture a swarm — unless you are certain this swarm originated from your hive.

Be prepared for a crowd of awestruck onlookers. I always draw a crowd when I capture a swarm. Everyone in your audience will be stunned as you walk up to this mass of 20,000 stingers wearing only a veil for protection. "Look," they'll gasp, "that beekeeper is in short sleeves and is not wearing any gloves! Is she crazy?" Only you will know the secret: The bees are at their gentlest when they're in a swarming cluster. You have nothing to fear. But your neophyte audience will think that your bravery is supreme. To them, you are a bee charmer — or the bravest (or nuttiest) person alive!

The easiest swarms to capture are those that are accommodating enough to collect on a bush or a low tree branch — one that you can reach without climbing a ladder. Obviously, if the branch is high up in a tree, you should not attempt your first capture! Gain experience by first capturing swarms that are easy reaches. Then you can graduate to the school of acrobatic swarm collection.

Say your swarm is located on an accessible branch. Lucky you! Follow these steps to capture it:

1. **Place a suitable container on the ground below the swarm.**

   You can use a large cardboard box (my favorite), an empty beehive, or a nuc box (see Chapter 5). This container will be the swarm's temporary accommodation while you transport the bees to their new, permanent home.

2. **Get the bees off the branch.**

   One approach is to give the branch holding the bees a sudden authoritative jolt. Doing so will dislodge the swarm, and it will (hopefully) fall into the container that you have placed directly under it. If this approach works, great. But it can be tricky. The swarm may miss its mark, and you may wind up with bees all over the place. In addition, this violent dislodging tests the gentle demeanor I promised!

   I prefer a more precise approach that enables you to place (not drop) the bees into their "swarm box." This approach works if the swarm is on a branch that you can easily sever from the rest of the foliage. You'll need a pair of pruning shears — a size appropriate for the job at hand. Follow these steps:

   1. **Study the swarm.**

      Notice how the bees are clustered on the branches. Can you spot the main branch that's holding the swarm? Are several branches holding it? Try to identify the branch (or branches) that, if severed, will allow you to gingerly walk the branch with swarm attached over to the box. In this manner, you can *place* the swarm in the box, not dump it.

   2. **Snip away at the lesser branches while firmly holding the branch containing the mother lode with your other hand.**

      Work with the precision of a surgeon: You don't want to jolt the swarm off the branch prematurely. When you're absolutely sure that you understand which branch is holding the bees, make the decisive cut. Anticipate that the swarm will be heavier than you imagined, and be sure that you have a firm grip on the branch before you make the cut. Avoid sudden jolts or drops that would knock the bees off the branch.

3. **Carefully walk the swarm (branch and all) to the empty cardboard box and place the whole deal in the box.**

4. **Close up the box, tape it shut, and you're done.**

   Get it home right away because heat will build up quickly in the closed box.

   I have modified a cardboard box for swarm captures. One side contains a "window" that I have fitted with mosquito screen. This window gives the captured swarm ample ventilation.

Hiving your swarm

You can introduce your swarm into a new hive in the following manner:

1. **Decide where you want to locate your new colony.**

   Keep in mind all the factors you need to consider when making this decision (see Chapter 3).

2. **Set up a new hive in this location.**

   You'll need a bottom board, a deep-hive body, ten frames and foundation, an inner cover, and an outer cover. Keep the entrance wide open (no entrance reducer).

3. **Place a bed sheet in front of the new hive, from the ground to the hive entrance.**

   This ramp will help the bees find the entrance to their new home. In lieu of a bed sheet, you can use a wooden plank or any configuration that creates a gangplank for the bees.

4. **Take the box containing the swarm and shake/pour the bees onto the bed sheet, as close to the entrance as possible.**

   Some of the bees will immediately begin fanning an orientation scent at the entrance, and the rest will scramble right into the hive. What a remarkable sight this is — thousands of bees marching into their new home.

The swarm of bees (now in their new home) will draw comb quickly because they arrive loaded with honey. You can use a hive-top feeder if you wish, but it may not be necessary given the time of year that swarms occur.

In a week, check the hive and see how the bees are doing. See any eggs? If you do, you know the queen is already at work. How many frames of foundation have been drawn into comb? The more the merrier! Is it time to add a second deep (see the "7/10 rule" mentioned earlier in this chapter)?

Absconding

Absconding doesn't happen very often, but it's a cruel blow when it does. One day, you go to the hive and find that no one's at home. Every last bee (or nearly every bee) has packed up and left town. What a horror! Here are some of the typical causes of absconding:

- **Lack of food.** Make sure that your hive has an ample supply of honey. Feed your bees sugar syrup when their stores are dangerously low and during serious dearths of nectar.

- **Loss of queen.** This situation eventually results in a hive with no brood. Always look for evidence of a queen when you inspect your bees. Look for eggs!

- **Uncomfortable living conditions.** Make sure that the hive is situated where it doesn't get too hot or too wet. Overheated or overly wet hives make life unbearable for the colony. Provide ample ventilation and tip the hive forward for good drainage.

- **Itty-bitty (or not so itty-bitty) pests.** Some hives (particularly weak ones) can become overrun with other insects, such as ants, hive beetles, or wax moths. Even persistent raids from wildlife (skunks, raccoons, and bears, for example) can make life miserable for the bees. See Chapter 11 for tips on dealing with these annoyances.

- **Mites and disease.** Colonies that are infested with mites or have succumbed to disease may give up and leave town. Routinely medicate your bees to prevent such problems (see Chapter 11).

Where Did the Queen Go?

It's every beekeeper's nightmare: The queen is dead, or gone, or lost. Whatever the reason, if the colony doesn't have a queen, it's doomed. That's why you must confirm that the queen is alive and well at every inspection. If you come to the dismal conclusion that your colony is queenless, you can do two things: let the colony raise its own queen or introduce a new queen into the colony.

Letting nature take its course

To let the colony create a new queen, it must have occupied queen cells or cells with eggs. If eggs are available, the worker bees will take some of them and start the remarkable process of raising a new queen. When the new virgin queen hatches, she will take her nuptial flight, mate with drones, and return to the hive to begin laying eggs. If no eggs are available for the colony to raise a new queen, you must take matters into your own hands and order a new queen from your beekeeping supplier (see next section).

The colony must have eggs to create its own queen. Older larvae or capped brood are at too late a developmental stage to be transformed into new queens.

Replacing your queen naturally is certainly interesting, but consider the logistics. The entire process (from egg to laying queen) can take a month. That's a precious amount of time during honey collection season. In the interest of productivity, it may be better to take matters into your own hands and order a replacement queen.

Ordering a replacement queen

A faster solution than the *au natural* method is to order a replacement queen from your bee supplier. Within a few days, a vigorous queen will arrive at your doorstep. She's already mated and ready to start producing brood.

The advantages of ordering a queen are clear:

✔ It provides a fast solution to the problem of having a queenless colony.

✔ The queen is certain to be fertile.

✔ It guarantees the pedigree of your stock. (Queens left to mate in the wild can produce bees with undesirable characteristics, such as a bad temper.)

Introducing a new queen to the hive

After your queen arrives by mail, you must introduce her into the colony. Doing so can be a little tricky. You can't just pop her in: She's a stranger to the colony, and it is sure to kill her. You have to introduce her *slowly*. The colony needs time to accept her and become accustomed to her scent. Old-time beekeepers swear by all kinds of methods — and some are downright weird. (I don't want you to try them so I'm not going to mention them here!) I suggest that you use the following approach:

1. **Remove one of the frames from the brood box.**

    Pick a frame with little or no brood on it, as whatever brood is on the frame will be lost — you won't use this frame again for a week.

2. **Shake all the bees off the frame and put it aside for the next week.**

3. **With the one frame removed, create a space in the center of the brood box. Use this space to hang the queen cage in the same way you hung it when you first installed your package bees (see Chapter 5 and Figure 9-4).**

TIP

Make sure to remove the cork from the queen cage to expose the candy plug. Also, when you hang the cage, make sure that the candy end is facing *up*. That way, any attendant bees that die in the cage will not block the hole and prevent the queen from getting out. Leave the bees alone for one week, and then inspect the hive to determine that the queen has been released and that she is laying eggs.

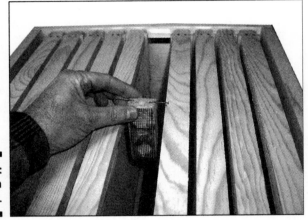

Figure 9-4:
Hanging a
queen cage.

If the weather is mild (over 60 degrees F at night), you can introduce the queen cage on the bottom board (see Figure 9-5). Remove the cork to expose the candy plug. Slide the cage screen side up along the bottom board and situate it toward the rear of the hive. Leave the bees alone for one week, and then inspect the hive to determine that the queen has been released and that she is laying eggs.

Figure 9-5:
Sliding a
queen cage
onto the
bottom
board.

Avoiding Chilled Brood

Honey bees keep their hive clean and sterile. If a bee dies, the others remove it immediately. If a larva or pupa dies, out it goes. During the early spring, the weather can be unstable. A cold weather snap can chill and kill some of the developing brood. When this happens, the bees dutifully remove the little corpses and drag them out of the hive. Sometimes the landing board at the entrance is as far as they can carry them. You may spot several dead brood at the entrance or on the ground in front of the hive. Don't be alarmed — the bees are doing their job. A few casualties during the early spring are normal.

Note: Chilled brood looks similar to, but is different from, the disease *chalkbrood.* You can find information about chalkbrood in Chapter 10.

Sometimes beekeepers unwittingly contribute to the problem of chilled brood. Remember, chilled brood is killed brood. You can do a few things to avoid endangering your bees:

- ✔ When the temperature drops below 50 degrees F, keep your inspections very, very brief. A lot of heat escapes every time you open the hive, and brood can become chilled and die.

- ✔ Provide adequate ventilation to avoid condensation. The resulting icy water dripping on the bees can chill the brood.

- ✔ Inspect your bees only on days when there is little or no wind (especially during cool weather). Harsh winds will chill brood.

Dealing with the Dreaded Robbing Frenzies

Robbing is a situation in which a hive is attacked by invaders from other hives. The situation is serious for a number of reasons:

- ✔ A hive defending itself against robbing will fight to the death. This battle can result in the loss of many little lives and even destroy an entire colony. Tragedy!

- ✔ If the hive is unable to defend itself in a robbing situation, the invading army can strip the colony of all its food. Disaster!

- ✔ Being robbed changes the disposition of a hive. The bees can become nasty, aggressive, and difficult to deal with. Ouch!

Many new beekeepers mistake a robbing situation as being the opposite of a problem. Look at all that activity around the hive! Business must be booming!

It's a natural mistake. The hive's entrance is furious with activity. Bees are everywhere. Thousands of them are darting in, out, and all around the hive. But look more closely. . . .

Knowing the difference between normal and abnormal (robbing) behavior

A busy hive during the nectar flow may have a lot of activity at the entrance, but the normal behavior of foraging bees looks different than a robbing situation. Foraging bees go to and fro with a purpose. They shoot straight out of the hive and are quickly up and away. Returning foragers are weighted down with nectar and pollen and land solidly when returning to their hive. Some even undershoot the entrance and crash-land just short of the bottom board.

Other times, normal activity at the hive's entrance can look unusually busy. This is when young worker bees take their orientation flights. Facing the hive, they hover up, down, and back and forth. They're orienting themselves to the location of their hive. You may see hundreds of these young bees floating around the front of the hive, but there's nothing aggressive or frantic about their exploratory behavior.

In contrast to these normal busy situations, robbing takes on an aggressive and sinister look. Learn to recognize the warning signs:

- ✔ Robbing bees approach the hive without being weighted down with nectar. They may not shoot right into the entrance. Instead, they fly from side to side, waiting for an opportune moment to sneak past the guard bees.

- ✔ If you look closely, you may see bees fighting at the entrance or on the ground in front of the hive. They are embraced in mortal combat. These are the guard bees defending their colony to the death.

- ✔ Unlike foraging bees that leave the hive empty-handed, robbing bees leave the hive heavily laden with honey, which makes flying difficult. Robbing bees tend to climb up the front of the hive before taking off. Once they're airborne, there's a characteristic dip in their flight path.

Putting a stop to a robbing attack

If you think that you have a robbing situation underway, don't waste time. Use one or more of the following suggestions to halt robbing and prevent disaster:

- ✔ Reduce the size of the entrance to the width of a single bee. Doing so will make it far easier for your bees to defend the colony. But be careful. If the temperature has turned hot, narrowing the entrance impairs ventilation.

✔ Soak a bed sheet in water and cover the hive that's under attack. The sheet draped to the ground prevents robbing bees from getting to the entrance. The bees in the hive seem to be able to find their way in and out. During hot, dry weather, rewet the sheet as needed. Be sure to remove the sheet after one or two days.

✔ A product on the market called Liquid Bee Smoker (see Figure 9-6) is a smoky-smelling concentrate that's diluted with water and put into a spray bottle. If your hive is under attack, spray all the bees at the entrance with Liquid Bee Smoker. The robbing frenzy should come to an instant halt.

I use Liquid Bee Smoker frequently for brief inspections (instead of firing up my smoker). I was thrilled to learn of another use for this product.

Figure 9-6:
Use Liquid Bee Smoker to end a robbing frenzy.

Preventing robbing in the first place

The best of all worlds is to prevent robbing from happening at all. Here's what you can do:

✔ Never leave honey out in the open where the bees can find it — particularly near the hive and during a dearth in the nectar flow. Easy pickings can set off a robbing situation.

✔ When harvesting honey, keep your supers covered after you remove them from the colony.

✔ Be very careful when handling sugar syrup. Try not to spill a drop when feeding your bees. The slightest amount anywhere but in the feeder can trigger disaster.

✔ Until your hive is strong enough to defend itself, use the entrance reducer to restrict the size of the opening the bees must protect. Also, be sure to close off the hole in the inner cover.

✔ Never feed your bees in the wide open (such as filling a dish with syrup or honey and putting it near the entrance of the hive).

✔ Avoid using an entrance feeder (see Figure 9-7). Being so close to the entrance, these feeders can incite robbing behavior.

Figure 9-7: I don't advocate using an entrance feeder — it can entice strange bees to rob your hive.

Don't be tempted to make it easier for your bees to access the syrup you feed them. I know of a beekeeper who put shims between the hive-top feeder and hive to create a gap that makes it easier for the bees to access the syrup. The result was a furious robbing attack from other bees. Keep your feeding device where only your colony can reach it.

Ridding Your Hive of the Laying Worker Phenomenon

If your colony loses its queen and is unable to raise a new queen, a strange situation can arise. Without the "queen substance" wafting its way through the hive, there is no pheromone to inhibit the development of the worker bees' reproductive organs. In time, young workers' ovaries begin to produce eggs. But these eggs are not fertile (the workers are incapable of mating). So the eggs can only hatch into drones. You may notice eggs, larvae, and brood and never suspect a problem. But you have a huge problem! In time, the colony will die off without a steady production of new worker bees to gather food and tend to the young. A colony of drones is doomed.

How to know if you have laying workers

Be on the lookout for a potential laying-workers situation and take action when it happens. The following are key indicators:

- ✔ **You have no queen.** Remember that every inspection starts with a check for a healthy, laying queen. If you have lost your queen, you must replace her.

- ✔ **You see lots and lots of drones.** A normal hive never has more than a few hundred drone bees. If you notice a big jump in the drone population you may have a problem.

- ✔ **You see cells with two or more eggs.** This is the definitive test. A queen bee will place only one egg in a cell — never more than one. Laying workers are not so particular; they will place two or more eggs in a single cell. If you see more than one egg in a cell (see Figure 9-8), you can be certain that you have laying worker bees. Time to take action!

Figure 9-8:
The best way to determine whether you have laying workers is to count eggs in the cells. If you spot multiple eggs in a cell, you have a problem to deal with.

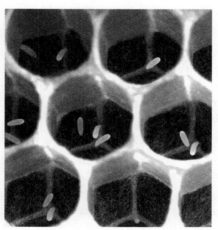

Getting rid of laying workers

You may think that introducing a young and productive queen will set things right. But it won't. The laying workers will not accept a queen once they have started laying eggs. If you attempt to introduce a queen, she will be swiftly killed. Guaranteed.

Before you can introduce a new queen, you need to get rid of all the laying workers. But how? They look just like all the other workers! The solution is tedious and time-consuming but 100 percent effective when done properly. You need the following items:

- An empty hive (no frames)
- An outer cover
- A wheelbarrow or hand truck

Follow these steps:

1. **Order a new marked queen from your bee supplier.**

2. **The day your queen arrives, put the entire hive (bees and all, minus the bottom board) in the wheelbarrow (or on the hand truck) and move it at least 100 yards away from its original location.**

   The bottom board stays in its original location.

3. **One by one, shake every last bee off each frame and onto the grass.**

   Not a single bee can remain on the frame — that bee might be a laying worker. A bee brush (see Chapter 4) helps get the stubborn ones off.

4. **Put each empty frame (without bees) into the spare hive you have standing by.**

   Make sure that no bees return to these empty frames. Use the extra cover to ensure that they can't get to these denuded frames.

5. **When you have removed the bees from every frame, return the frames to the original hive bodies.**

   Again, just make sure that no bees sneak back onto the frames.

6. **Roll the hive and empty frames back to their original location and place the hive on the bottom board.**

   Some of the bees will be there waiting for you. These are the older foraging bees (not the younger laying workers). Be careful not to squash any bees as you slide the hive back onto the bottom board.

Most of the older foraging bees will find their way back to the hive. But the young nurse bees, the ones that have been laying eggs, have never ventured out of the hive before. They will be lost in the grass where you deposited them and will never find their way back to the hive.

Now you can safely introduce your new queen. See the instructions earlier in this chapter.

Preventing Pesticide Poisoning

I get upset when I see people spraying their lawns and trees with pesticides. These chemicals may make for showcase lawns and specimen foliage, but they can't be good for the water table, birds, and other critters. And I know some of these treatments can be deadly for bees. (**Note:** I'm not talking about fertilizers here, just pesticides.) If you ever see a huge pile of dead bees in front of your hive, you can be pretty sure that your girls were the victims of pesticide poisoning. Here are a few things you can do to avoid such a tragedy:

✔ Let your neighbors know that you are keeping bees. Make sure that they know how beneficial pollinating bees are to the community and ecology. Explain to them the devastating effect that pesticide spraying can have on a colony. They may think twice about doing it at all. If they must spray, urge them to do so at dawn or dusk, when the bees are not foraging. Encourage your neighbors to call you the day before they plan to spray. With advance warning, you can protect your bees.

✔ On the day your neighbors plan to spray, cover your hive with a bed sheet that you have saturated with water. Let it drape to the ground. The sheet will minimize the number of bees that fly that day. Remove the sheet the following morning after the danger has passed.

✔ Register your colony with the state apicultural office or agricultural experiment station. You may have to pay a minimal charge for registration. Each state publishes a list of all registered beekeepers in the state. Reputable arborists check such lists before spraying in a community. If you are on the list, they will call you before they spray in your area.

The Killer Bee Phenomenon

The media has had a ball with the so-called *killer bees.* These nasty-tempered bees have been fodder for fantastic headlines and low-budget horror movies. At the same time, this kind of publicity has had a negative and unwelcome impact on backyard beekeeping. The resulting fear in the community can make it difficult for a beekeeper to gain the support and acceptance of his or her neighbors. Moreover, sensational headlines have resulted in sensational legislation against keeping bees in some communities. The public has been put on guard. Killer bees present another problem for the beekeeper as well: If your area has them, you must manage your colony carefully to prevent your own bees from hybridizing and becoming more aggressive.

A "bee" movie

Hollywood producer Irwin Allen is the king of the disaster movies. In his movie *The Swarm,* great clouds of "killer bees" attack entire cities and leave hundreds dead in their wake. It was a disaster movie in more ways than one. As a business venture, it tanked at the box office. As a public relations vehicle for beekeepers, it fueled a fire of fear in the minds of the public.

What are "killer bees"?

First of all, let's get the name correct. The bees with the bad PR are actually Africanized Honey Bees (AHB) — or *Apis mellifera scutellata* if you want to get technical. The "killer bee" pseudonym was the doing of our friends in the media.

How did the AHB problem come about? It all started in 1956 in Brazil. A group of scientists were experimenting with breeding a new hybrid resulting in superior honey production. They were breeding the notoriously aggressive honey bee from Africa with the far more docile European honey bee. But a little accident happened. Some African queen bees escaped into the jungles of Brazil. The testy queens interbred with European bees in the area, and voilà — the AHB become a force to deal with.

Outwardly, AHBs look just like European honey bees. In fact, you must take a peek under the microscope to detect the difference. Their venom is no more powerful. And like our sweet bees, they too die after inflicting a sting. The main and most infamous difference is their temperament. They are very defensive of their hives, are quick to attack, will chase an intruder long distances, and stay angry for days after an incident.

There have been reports of human deaths resulting from attacks by AHBs. But these reports are rare and almost always involve elderly victims who have been unable to fend off the attackers. The media can put quite a sensational spin on such tragedies, and that has contributed to some bad PR for honey bees in general.

My friend Kate Solomon worked for several years in the Peace Corps teaching South American beekeepers how to work with the AHB. Kate's memorable "bee-beard" stunt (see Figure 9-9) resulted in not a single sting from these "killer bees." And yes, she has cotton in her nose and ears to keep unwanted explorers at bay!

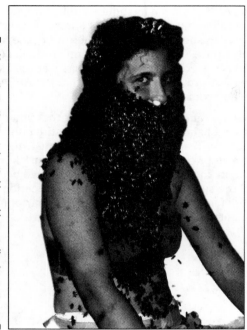

Figure 9-9:
AHBs are not always what the media has hyped them up to be. But for goodness sake, don't try such a stunt yourself! AHBs are capable of being very aggressive and dangerous.

Bee prepared!

In the nearly five decades since "the accident" in Brazil, AHBs have been making their way northward to the United States. A few years ago, the first colonies of AHBs were identified in some southern states. There is plenty of speculation as to how far north these bees are capable of surviving (after all, they are a tropical species). In any event, they have arrived amid great publicity. Beekeepers and the public will have to learn how to deal with them.

Here are some helpful hints about safe beekeeping in areas known to be populated by AHBs:

✔ If you live in an area where AHBs have been seen, do not capture swarms or populate your hive with anything other than package bees from a reputable supplier. Otherwise, you may wind up with the hive from hell.

✔ If you are unlucky enough to disturb a colony of AHBs, don't stick around to see how many will sting you. Run in a straight line far away from the bees. AHB are fast flyers, and you will have your work cut out for you when you attempt to outrun them. Don't jump into water — they'll be waiting for you when you surface. Instead, enter a building and stay inside until things cool off.

✔ In the areas where the AHB has been introduced, diligent beekeepers are the community's best defense against the AHBs spread. By systematically inspecting her hive to spot her *marked* queen, a beekeeper knows that her colony remains pure. Only when an unfamiliar queen (perhaps an AHB) is introduced is the colony's genetic integrity at risk. More than ever, backyard beekeepers are needed to ensure that the AHB doesn't become a problem in any community.

✔ If you join a local bee club (and I highly recommend that you do), encourage the club to publish information educating the public about the benefits of beekeeping. Teach the community the real story about the AHB. Take positive steps to quell the fear that may lurk in some people's minds. Let them know how important it is to have beekeepers who can help control the spread of the AHB. A good education program is a beekeeper's best defense against local legislation restricting beekeeping in the community.

Chapter 10

Diseases and Remedies

· ·

In This Chapter:

▶ Deciding to medicate

▶ Preventing problems before they happen

▶ Recognizing the first signs of trouble

▶ Nursing sick bees back to good health

· ·

1 won't pretend otherwise — this is *not* the fun part of beekeeping. I'd much rather never have to think about my bees getting sick. My heart aches when they do. Nothing is more devastating than losing a colony to disease. But let's get real. Honey bees, like any other living creatures, are susceptible to illness. Although some of these diseases aren't too serious, some can be devastating. The good news is that you can prevent many honey bee health problems *before* they happen, and you can often head off disaster if you know the early signs of trouble.

Right away let me clear up one thing. None of the health problems that affect bees have any impact on human health. These diseases are 100 percent unique to your bees. They're not harmful or contagious in any way to you or your family. Phew! That's a relief!

In this chapter I've highlighted the most common health problems that your bees may face. As you inspect your hives, look carefully at the capped and open brood cells (what's going on in these cells is often the barometer of your colony's health). Discover how to recognize the telltale indications of health problems.

Medicating, Or Not?

I know what you're thinking. Should you put medication in your hive or not? Wouldn't keeping everything natural and avoiding the use of any chemicals, medications, or antibiotics be better? Maybe you can even save a few dollars? Well, perhaps the answer to that question depends on your practice in other areas. Do you avoid taking your dog to the vet for distemper shots and heartworm pills? Do you send your kids off to school without their vaccines?

Probably not. Bees are no different. Without some help from you, I can assure you they'll eventually get sick. You may even run the risk of losing your hive entirely. Don't risk it. Follow a sensible annual medication regime and look carefully for signs of trouble every time you inspect your colony.

Remember that you should never ever medicate your bees when you have honey on the hive that is intended for human consumption. In the early spring, medicate *before* honey supers go on the hive. In the autumn, medicate *after* they are removed. For a description of honey supers and their use, see Chapter 4.

Knowing the Big Seven Bee Diseases

You should be on the lookout for seven honey bee diseases. Some are rare and it's doubtful that you'll ever encounter them. Some are more commonplace (like nosema and chalkbrood) and knowing what to do if they come knocking is important. Some, like American foulbrood, are quite serious, and you need to know how to recognize and deal with them.

Each time you inspect your bees, you're looking for two things: evidence of the queen (look for eggs) and evidence of health problems (look for the symptoms I describe below).

American foulbrood (AFB)

American foulbrood (AFB) is a nasty bacterial disease that attacks larvae and pupae. This serious threat is highly contagious to bees (not people) and, left unchecked, is certain to kill your entire colony. It's the worst of the bee diseases. Some symptoms are

- Infected larvae change color from a healthy pearly white to dark brown and die *after* they're capped.
- Cappings of dead brood sink inward (becoming concave) and often appear perforated with tiny holes.
- The capped brood pattern no longer is compact, but becomes spotty and random (see "Spotting Problems" in this book's color-photo section).
- The surface of the cappings may appear wet or greasy.

If you see these conditions, check for AFB by thrusting a toothpick or matchstick into the dead brood, mixing it around, and then *slowly* withdrawing the toothpick. Observe the material that is being drawn out of the cell as you withdraw the toothpick. Brood killed by AFB will be stringy and will *rope out* about ¼ inch (like pulling taffy) and then snap back like a rubber band (see

"Spotting Problems" in this book's color-photo section). Some dead pupae may have tongues protruded at a right angle to the cell wall. There may also be a telltale odor associated with this disease. Most describe it as an unpleasant "foul" smell (like a pot of old fashioned horse glue). If you suspect a foul smell, and that smell lingers in your nose after leaving the hive, your bees might have AFB.

Feeding your bees the antibiotic Terramycin in the spring prevents AFB (see Chapter 8). But if you suspect that your bees actually *have* AFB, immediately ask your state bee inspector to check your diagnosis. Treatment for AFB is subject to state law in the United States. If AFB is present, it is likely that your hives and equipment will have to be burned and destroyed. Why such drastic measures? Sleeping spores of AFB can remain active (even on old unused equipment) for up to 70 years.

Be wary of purchasing old, used equipment. If the former beekeeper's bees ever had AFB, the disease-causing spores remain in the equipment. No amount of scrubbing, washing, or cleaning can remedy the situation.

European foulbrood (EFB)

European foulbrood (EFB) is a bacterial disease of larvae. Unlike AFB, larvae infected with EFB die *before* they're capped. Symptoms of EFB include the following:

- ✔ Infected larvae are twisted in the bottoms of their cells like an inverted corkscrew. The larvae are either a light tan or brown color, and have a smooth "melted" appearance (see "Spotting Problems" in this book's color-photo section). Remember that normal, healthy larvae are a glistening bright white color.

- ✔ With EFB, many larvae die in their cells before they are capped. This makes it easy for you to see the discolored larvae.

- ✔ Capped cells may be sunken in and perforated, but the "toothpick test" won't result in the telltale ropy trail as described above for AFB.

- ✔ A sour odor may be present (but not as foul as AFB).

Here's the best way to view frames for diseased larvae. Hold the frame by the ends of the top bar. Stand with your back to the sun and the light shining over your shoulder and down into the cells. The frame should be sharply angled so you are looking at the true *bottom* of the cell. Most new beekeepers interpret the "bottom" as the midrib of the comb. It isn't. The true bottom of the cell is the lower wall of the cell (the wall that's closest to the hive's bottom board when the frame is hanging in the hive).

Keep your eye on Tylosin

Terramycin is currently the only antibiotic approved for use with honey bees. For years it has been the beekeepers best defense against American foulbrood. But in recent years, some strains of American foulbrood have become resistant to this antibiotic. So the search is on for a new and more effective treatment. The answer may be Tylosin — an alternate antibiotic. It is currently approved for use on some farm animals, but it is not yet approved for use on honey bees. But that may change. So keep your eye on this product. It may well become a welcome addition to the beekeeper's arsenal of medication.

Because EFB bacteria don't form persistent spores, this disease isn't as dangerous as AFB. Colonies with EFB sometimes recover by themselves after a good nectar flow begins. Although serious, EFB is *not* as devastating as AFB and can be successfully treated with antibiotics.

Treating colonies in the spring with Terramycin prevents or controls EFB (see Chapter 9). If you've detected EFB, requeen your colony (replace the old queen with a new one; see Chapter 9) to break the brood cycle and allow the colony time to remove infected larvae. Help the bees out, and remove as many of the infected larvae as you can using a pair of tweezers.

Nosema

Nosema, a common protozoan disease that affects the intestinal tracks of adult bees is kind of like dysentery in humans. It can weaken a hive and reduce honey production by between 40 and 50 percent. It's most common in spring after bees have been confined to the hive during the winter. Some symptoms of nosema are

- ✔ In the spring, infected colonies build up slowly or perhaps not at all.
- ✔ Bees appear weak and may shiver and crawl aimlessly around the front of the hive.
- ✔ The hive has a characteristic *spotting,* which refers to streaks of mustard-brown feces that appear in and on the hive.

You can discourage nosema by selecting hive sites that have good airflow and a nearby source of fresh, clean water. Avoid damp, cold conditions that can encourage nosema. Creating an upper entrance for the bees during winter improves ventilation and discourages nosema.

Treat nosema by feeding fumagillin (Fumidil B) in sugar syrup in spring and fall. See Chapter 8 for how to prepare and feed medicated sugar syrups.

Chalkbrood

Chalkbrood is a common fungal disease that affects bee larvae. Chalkbrood pops up most frequently during damp conditions in early spring. It is rather common and not that serious. Infected larvae turn a chalky white color, become hard, and may occasionally turn black. You may not even know that your bees have it until you spot the chalky carcasses on the hive's "front porch." Worker bees on "undertaker duty" attempt to remove the chalkbrood as quickly as possible, often dropping their heavy loads at the entrance or on the ground in front of the hive (see "Spotting Problems" in this book's color-photo section).

Misdiagnosing this disease is common, because it's easily confused with chilled brood (see Chapter 9). You see carcasses at the hive entrance with both anomalies, but with chalkbrood, the bodies are hard and chalky (not soft and translucent as is true with chilled brood).

No medical treatment is necessary for chalkbrood. Your colony should recover okay on its own. But you can help them out by removing mummified carcasses from the hive's entrance and from the ground around the hive. Also, usually one frame will have most of the chalkbrood cells. Remove this frame from the hive and replace it with a new frame and foundation. This action minimizes the bees' job of cleaning up. Your help quickly arrests the spread of the fungus.

Sacbrood

Sacbrood is a viral disease of brood similar to a common cold. It isn't considered a serious threat to the colony. Infected larvae turn yellow and eventually dark brown. They're easily removed from their cells, because they appear to be in a water-filled sack. Now you know where the name comes from.

When to add Fumidil-B

You should add Fumidil B to the first 2 gallons of sugar syrup that you feed your bees in the spring. Also medicate the first two gallons you feed them in the autumn (see Chapter 8). Any additional gallons of syrup you feed to the bees are not medicated.

Honey bee viruses

Adult honey bees may occasionally fall prey to various different kinds of viruses. Viruses aren't easily detected and are often overlooked by beekeepers. Perhaps the most easily recognized virus is *chronic bee paralysis,* which causes workers to become greasy looking, hairless, and uniformly black in color. Sick bees are seen crawling on the grass in front of the hive, simply unable to fly. ***Note:*** Colonies infested with mites (see Chapter 11) are far more susceptible to viral diseases, because open wounds created by mites are an invitation to infection.

No medical treatment exists for honey bee viruses, but you can help. One by one, remove each frame from the hive and carry it 10-20 feet away. Now shake all the bees off the frame and return it (empty) to the hive. Do this for all frames. The sick bees will not be able to return to the hive. The healthy ones will have no trouble making it home.

No recommended medical treatment exists for sacbrood. But you can shorten the duration of this condition by removing the sacs with a pair of tweezers. Other than that intervention, let the bees slug it out for themselves.

Do your best to keep your bees free of stressful problems (mites, poor ventilation, crowded conditions) and they'll have an easier time staying healthy and avoiding disease.

Stonebrood

Stonebrood is a fungal disease that affects larvae and pupae. It is rare and doesn't often show up. Stonebrood causes the mummification of brood. Mummies are hard and solid (not sponge-like and chalky as with chalkbrood). Some brood may become covered with a powdery green fungus.

No medical treatment is recommended for stonebrood. In most instances worker bees remove dead brood and the colony recovers on its own. You can help things along by cleaning up mummies at the entrance and around the hive, and removing heavily infested frames (see treatment for chalkbrood).

A handy chart

Table 10-1 gives you a quick overview of the big seven bee diseases, their causes, and their distinguishable symptoms. It contains a description of a healthy bee colony for comparison purposes.

Table 10-1   **Honey Bee Health at a Glance**

Situation/Disease	What Causes It?	Appearance of Brood	Appearance of Cappings	Appearance of Dead Larvae	Color and Consistency of Larvae	How Does it Smell?
Normal, healthy brood and bees	The result of terrific beekeeping!	Tight pattern of sealed and open brood cells	Light tan, brown color, slightly convex	No dead larvae	Plump, bright white, wet, pearly appearance	Fresh sweet smell (or no smell at all)
American foulbrood (AFB)	A bacterium (spore forming)	Scattered, spotty brood pattern	Sunken, perforated, discolored, greasy appearance up to the roof of the cell.	Flat and fluid-like on bottom of cell. Tongue extended	Brown, dull, sticky and ropy	Unpleasant, sharp, foul smell
Chalkbrood	A fungus	Scattered, spotty brood pattern	Sunken, perforated, discolored	Most often in sealed or perforated cells	White and moldy, later white, gray or black, hard and chalk-like	Normal
European foulbrood (early stages)	A bacterium	Scattered, spotty brood pattern	Some discolored, sunken, perforated	In unsealed cells, in twisted positions	Yellowish, tan or brown	Sour
European foulbrood (advanced stage)	A bacterium	Scattered, spotty brood pattern	Discolored, sunken, perforated	In unsealed and sealed cells, in twisted positions	Brown, but not ropy or sticky	Sour
Sacbrood	A virus	Scattered brood pattern. Many unsealed cells	Often dark and sunken, many perforated	Most often with head raised	Grayish to black, skin has a watery, sack-like appearance	Sour or no smell
Stonebrood	A fungus	Affected brood are usually white, but can sometimes have a greenish, moldy appearance	Some cappings are perforated and covered with a greenish mold	In unsealed and sealed cells	Green-yellow, or white, hard and shrunken	Moldy

Chapter 11

Honey Bee Pests

• •

In This Chapter

▶ Learning about some common pests

▶ Recognizing and preventing potential problems

▶ Treating your colony when the going gets tough

▶ Utilizing some holistic alternatives to modern medicine

▶ Keeping out some furry nonfriends

• •

*E*ven healthy bee colonies can run into trouble every now and then. Critters (four-legged and multilegged) can create problems for your hives. Anticipating such trouble can head off disaster. And if any of these pests get the better of your colony, you'll need to know what steps to take to prevent things from getting worse.

In this chapter I introduce you to a few of the most common pests of the honey bee, and what you must do to prevent catastrophe.

Parasitic Mites

Two little mites have gotten a lot of publicity in recent years about big problems they've created for honey bees: the varroa mite and the tracheal mite. These parasites have become unwelcome facts of life for beekeepers, changing the way they care for their bees. By some accounts, up to 90 percent of the *feral* (wild) hives have died off as a result of these mites. Feral bees don't have beekeepers helping them fend off mites. You need to be aware of these pests and find out how to control them. Doing *nothing* to protect your bees from mites is like playing a game of Russian roulette.

Varroa mites

Somehow this little pest (*Varroa jacobsoni*) has made its way from Asia to all parts of the world, with the exception of Hawaii and Australia. Varroa has been in the United States since the late 1980s (maybe longer) and has created quite a problem for beekeepers. Resembling a small tick, this mite is about the size of a pinhead and is visible to the naked eye (see Figure 11-1 and the color-photo section of this book). Like a tick, the adult female mite attaches herself to a bee and feeds on its blood (*hemolymph fluid*).

Hemolymph fluid is the "blood" of arthropods. It is the fluid that circulates in the body cavity of an insect, and carries oxygen like mammal blood does.

Mites attached to foraging worker bees enable the infestation to spread from one hive to another. The varroa mite is strongly attracted to the scent of drone larvae, but it also invades other brood cells just before they're capped over by the bees. Within the cells varroa mites feed on the developing bees and lay eggs. They reproduce at a fantastic rate and cause a great deal of stress to the colony. The health of the colony can weaken to a point that bees become highly susceptible to viruses. Within a couple of seasons, the entire colony can be wiped out.

Figure 11-1:
Varroa mites can seriously weaken a hive by attaching to bees and feeding on their hemolymph (blood).

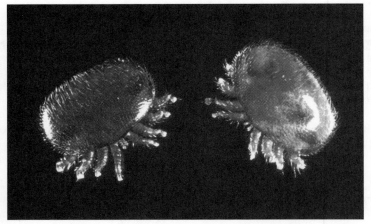

Courtesy of United States Department of Agriculture

Recognizing varroa mite symptoms

How do you know if your colony has a serious infestation of varroa mites? Following is a list of some varroa mite symptoms. If you suspect a varroa infestation, confirm your diagnosis using one of the surefire detection techniques I describe in the next section.

- ✔ Do you see brown or reddish spots on the white larvae? You may be seeing mites.

- ✔ Are any of the newly emerged bees badly deformed? You may notice some bees with stunted abdomens and deformed wings.

- ✔ Do you actually *see* varroa on adult bees? They're usually found behind the head or nestled between the bee's abdominal segments.

✔ Finding mites on adult bees indicates a *heavy* infestation. The mites head for bee larvae first (before the larvae are capped and develop into pupae). They then feed on capped pupae. It doesn't take much to figure out that by the time the mites are prevalent on adult bees, the mite population is quite high.

✔ Did your colony suddenly die in late autumn? Oops! You're way too late to solve the problem this year. You'll have to start fresh with a new colony next spring.

Utilizing two surefire detection techniques for varroa

If you suspect a varroa mite problem, then, by all means, confirm your diagnosis by using either the powdered sugar shake method or the drone brood inspection method. But performing one of these detection techniques before you suspect a problem is best. Varroa detection needs to be a routine part of your inspection schedule. I suggest using the powdered sugar shake method twice a year — once in the early spring, and once in the late summer.

Powdered sugar shake method

The powdered sugar shake technique is my favorite method for detecting varroa. It is effective and nondestructive (no bees are killed in the process). You use this process in the early spring (before honey supers go on) and again in the late summer (before the honey supers come off). Follow these steps:

1. **Obtain a wide-mouthed glass jar (the kind mayonnaise comes in) and modify the lid so that it has a coarse screen insert. Just cut out the center of the lid and tape or glue a wire screen over the opening (see Figure 11-2).**

   Hardware cloth (eight wires to the inch) works well. Now you have something resembling a jumbo saltshaker.

2. **Put 3 to 4 tablespoons of powdered sugar (confectioners' sugar) into the jar.**

3. **Scoop up about 100 to 200 live bees from the brood nest and place them in the jar. Be careful that you don't scoop up the queen! Screw on the perforated lid.**

4. **Cover the screened lid with one hand (to keep the powdered sugar from spilling out) and shake the jar vigorously (like a bartender making a martini).**

   This action doesn't really harm the bees, but it sure wakes them up!

5. **Shake the powdered sugar through the screened top and onto a white sheet of paper.**

   Shake authoritatively. Doing so dislodges any mites that are on the bees. The mites can easily be counted, contrasted against the white paper and powdered sugar.

Figure 11-2:
This jar's lid
has been
modified
for a
*powdered
sugar shake*
mite
inspection.

If you count one or more mites, you must proceed with the recommended treatment (see the "Knowing how to control varroa mite problems" later in this chapter). Seeing dozens of mites means the infestation has become significant. Take remedial action fast!

Note: Bees can be returned unharmed to the hive using this technique. Although they may be coated with sugar, their sisters nevertheless have a grand time licking them clean. Just wait 10 to 15 minutes to let them calm down before releasing them. All that jostling can make them understandably irritable.

Drone brood inspection method

Regrettably, the drone brood inspection method kills some of the drone brood. I prefer the sugar shake method for that reason alone. Follow these steps:

1. **Find a frame with a large patch of capped *drone* brood.**

   They are the larger capped brood with slightly dome-shaped cappings. Shake all the bees off the frame, and move to an area away from the hive where you can work undisturbed.

2. **Using an uncapping fork (see Chapter 12), slide the prongs along the cappings spearing the top third of the cappings and impaling the drone pupae as you shovel across the frame (see Figure 11-3).**

3. **Pull the drone pupae straight out of their cells.**

   Any mites are clearly visible against the white pupae (see Figure 11-4). Repeat the process to take a larger sampling.

Two or more mites on a single pupa indicate a serious, heavy infestation. Two or three mites per 50 pupae indicate a low to moderate infestation. But remember, whenever you see *any* mites at all, it's time to take action! (See "Knowing how to control varroa mite problems" next.)

Figure 11-3:
Slide the prongs of an uncapping fork along the drone brood cappings so you can check the pupae for varroa mites.

Courtesy of Stephen McDaniel

Figure 11-4:
Varroa mites first attach themselves to drone pupae, so that's a good place to look for evidence of an infestation. Can you see the mites on these pupae?

Courtesy of Stephen McDaniel

Knowing how to control varroa mite problems

A number of products and techniques are available that help reduce or even eliminate varroa mites populations. Here are the ones that I suggest you consider and a few that I think you should avoid as a new beekeeper.

Treating with Apistan

An effective and approved *miticide* (chemical that kills mites) is fluvalinate, which is sold under the brand name, *Apistan,* and is available from your bee-keeping supplier (see Figure 11-5). When any of the detection techniques mentioned earlier in this chapter indicate varroa mites, you must immediately treat with Apistan by carefully following the directions on the package.

Regardless of whether you detect varroa mites, I suggest that preventative miticide treatments be administered to your hive once in the spring and again in autumn.

Apistan is packaged as chemical-impregnated strips that look kind of like bookmarks. Hang two of the plastic strips in the brood chamber between second and third frames and the seventh and eighth frames (see Figure 11-6). You're positioning the strips close to the brood so the bees naturally come into contact with the miticide they contain.

Following package directions precisely is important when using Apistan in your hive. Remove honey supers before hanging Apistan strips and don't replace the supers until after the strips are removed at the end of the control period (six to eight weeks). Always wear gloves when handling the strips and make certain that each strip is placed between the frames, not on top of them. Use Apistan when daytime high temperatures are at least 50 degrees F. Don't reuse strips.

Leaving the product in the hive for too short or too long a time renders the treatment ineffective. For effective varroa control, keep strips in the hive for six to eight weeks. Don't remove strips from the hive for at least 42 days (six weeks). Do not leave strips in the hive for more than 56 days (eight weeks). Remove strips after six to eight weeks. Never leave Apistan in the hive over the winter. If you do, mites can build up a tolerance to the product, rendering it completely ineffective.

Figure 11-5: Apistan (the brand name for fluvalinate) effectively treats bees for varroa mites.

Figure 11-6:
Hanging
Apistan
strips.

Courtesy of Wellmark International

Never treat your bees with *any* kind of medication when you have honey supers on the hive. If you do, your honey becomes contaminated and cannot be used for human consumption. **Note:** Feeding medicated honey to the bees is, however, perfectly okay.

CheckMite+

Some mites have developed a resistance to Apistan, so new miticides have entered the market. CheckMite+ is a product manufactured by the Bayer Corporation (of aspirin fame). Like Apistan, it also consists of strips impregnated with a chemical miticide. But in the case of CheckMite+ the chemical is *coumaphos* — an ingredient used in deadly nerve gas. The product has not been approved for use in all states, so you need to check with your beekeeping supply dealer to determine whether you can even order it. In any event, it's tricky to use safely, and I don't recommend that you mess around with this product. My advice? New beekeepers should steer clear of CheckMite+.

Formic acid (Apicure)

Apicure, or formic acid, is available in gel packs, but it is so caustic and tricky to administer that I don't suggest that new beekeepers use it, either. Like CheckMite+, it also may not even be approved for use in your state. And it tends to result in a high mortality rate for queen bees.

Screened bottom board

About 10 percent to 15 percent of varroa mites routinely fall off the bees and drop to the bottom board. Under normal circumstances, they simply crawl back up into the hive, infesting the colony again. But that isn't possible when you use a screened bottom board (see Figure 11-7). The mites fall through the screen and straight to the ground.

The best screened bottom board designs come with a removable inspection sheet. When this sheet is in place, mites fall through the screen and become stuck to the sheet (you apply a thin film of petroleum jelly to the sheet to help the mites stick). Remove the sticky sheet to look for mites and determine infestation levels.

Using a screened bottom board is a natural (nonchemical) method for ridding the colony of up to 15 percent of an existing mite population. I urge you to consider replacing your regular bottom board with a screened one that has a removable tray (the tray can be put back into place in the winter to prevent drafts). In any event, a screened bottom board is an excellent way to improve ventilation in the hive.

Figure 11-7:
Using a
screened
bottom
board is
a useful
way to
detect
varroa
mites and
a non-
chemical
method for
controlling
mite
populations.

Tracheal mites

Another mite that can create serious trouble for your bees is the tracheal mite *(Acarapis woodi)* shown in Figure 11-8. These little pests are smaller than the period at the end of this sentence and can't be seen with the naked eye. Dissecting an adult bee and examining its trachea under magnification is the only way to identify a tracheal mite infestation.

As its name implies, this mite lives most of its life within the bee's trachea (breathing tubes), as shown in Figure 11-9. Mated female mites pass from one bee to another when the bees come in close contact with each other. Once the mite finds a newly emerged bee, she attaches to the young host and enters its tracheal tubes through one of the bee's spiracles — holes that are part of the respiratory system. Within the trachea the mite lays eggs and raises a new generation. The tracheal mite causes what once was referred to as *acarine disease* of the honey bee (a rather old fashioned term not used much these days).

In my opinion, this mite causes more trouble for hobbyist beekeepers than varroa. Early detection of bad infestations is difficult. As a result, tracheal mites can lead to the total loss of a colony before you're even aware that your bees are infested. Infestations are at their worst during winter months when bees are less active. Her majesty isn't laying eggs, so no new bees are emerging to make up for attrition. Winter also is when beekeepers don't routinely inspect the colony. Thus, seemingly healthy colonies with plenty of food sometimes suddenly die during late winter or early spring.

Symptoms that may indicate tracheal mites

The only surefire way to detect tracheal mites involves dissecting a bee under a microscope — a little tricky for the novice and not everyone has a dissection microscope in the hall closet. Whenever you suspect tracheal mites, call your state apiary inspector for information about how to have your bees inspected for tracheal mites.

A few clues may indicate the presence of tracheal mites. But the symptoms, listed below, are unreliable because they also may indicate other problems.

- You see many weak bees stumbling around on the ground in front of the hive. (This condition could also be an indication of Nosema disease; see Chapter 10.)

- You spot some bees climbing up a stalk of grass to fly, but, instead, then they just fall to the ground. This happens because mites clog the trachea and deprive the bee of oxygen to its wing muscles.

- You notice bees with *K-wings* (wings extended at odd angles — not folded in the normal position). This also can be an indication of Nosema disease.

- Bees abandon the hive (abscond) in early spring despite ample honey supplies. This can happen even late in the fall when it's too late to remedy the situation and making the time right for ordering package bees and starting anew in the spring.

How to control tracheal mite problems

Tracheal mite infestations are a problem, not a hopeless fate. You can take steps to use a number of techniques that I've listed in the following sections to prevent things from getting out of control. It isn't a case of just one technique working well. Play it safe by using a combination of some or all of these methods.

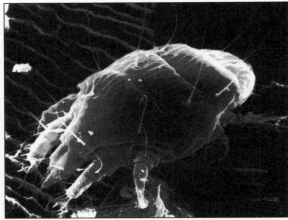

Figure 11-8: An adult tracheal mite (*Acarapis woodi*).

Courtesy of United States Department of Agriculture

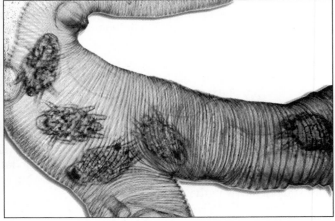

Figure 11-9: Tracheal mites (seen in this magnified photo of an infected bee's tracheal tubes) are responsible for *acarine disease,* a serious threat to the health of bee colonies.

Courtesy of United States Department of Agriculture

Menthol Crystals

Menthol crystals are the same ingredient found in candies and cough drops. Menthol is derived from a plant, making it a natural alternative to chemical miticides. Prepackaged bags containing 1.8 ounces of menthol crystals are available from your beekeeping supplier.

Place a single packet on the top bars of the brood chamber toward the rear of the hive (see Figure 11-10). Setting the packet on a small piece of aluminum foil prevents the bees from chewing holes in the bag and carrying away the menthol. Bees are tidy and try their best to remove anything they don't think belongs in the hive. Leave the menthol in the hive for 14 *consecutive* days when the outdoor temperature is between 60 and 80 degrees F. The menthol vapors are effective only at these temperatures. That means the product is *temperature dependent* — you can only use it when the weather is warm.

Honey for human consumption must be taken off the hive whenever *any* medications are used. You can safely apply honey supers three to four weeks after medication is removed from the hive.

Figure 11-10: In warm weather a packet of menthol placed on the top bars of the brood chamber helps control tracheal mites.

Menthol & oil saturated paper towel

The so-called *paper towel method* is an effective menthol treatment that isn't temperature dependent — it can be used year-round (except when honey for human consumption is on the hive). Follow these steps:

1. **Gently warm one pint of canola oil on the stove.**

2. **Once the oil is warm to the touch (not hot), dissolve the contents of one 1.8-ounce menthol packet in the oil.**

3. **Put the menthol-and-oil mixture in a one-gallon resealable plastic food bag, or in a shallow food storage container (the kind with an airtight snap-on lid) and place folded paper towels in the mixture.**

   Put as many folded towels in the oil as you can fit. You want the towels well saturated with the menthol-and-oil mixture.

4. **Place one menthol-oil paper towel on the top bars of the brood chamber and replace as needed.**

   Replace towels every two weeks during the winter when the mite does its worst damage. Keep unused paper towels tightly sealed in the plastic food container.

   The menthol eventually recrystallizes, but a short stint in the microwave (on half power) liquefies the mixture.

I suggest that you make this treatment part of your annual preventative plan (regardless of whether you see indications of having tracheal mites).

Sugar & grease patties

Placing patties of sugar and grease in the hive is a holistic treatment for tracheal mites that you can (and should) use year-round (even during the honey harvest season — unless you are adding the wintergreen oil option). As the bees feed on the sugar, they become coated with grease. The grease impairs the mite's ability to reproduce or latch onto the bees' hairs. Whatever the scientific reason, the treatment works effectively and is your number-one defense against tracheal mites.

Place one patty on the top bars of the brood chamber, flattening out the patty as needed to provide clearance for the inner cover and replacing it as the bees consume it. One patty should last a month or more.

Winter menthol-oil treatments

New beekeepers frequently ask me if they should apply menthol-and-oil paper towels throughout the winter and at what point they should stop the applications. By all means, use the menthol and oil paper towel method *throughout* the winter. It helps control tracheal mites when they could do the greatest damage.

Once a month during the winter (on a sunny day with no wind) take a quick peek under the inner cover. Have replacement towels ready to go and move briskly so very little heat will be lost when you open the hive. If you smoke the hive, do so very lightly when opening the hive during winter. You don't want to disturb the winter cluster.

Winter patty replacement

Feed your bees sugar & grease patties all year round (they help control tracheal mites). These patties have no effect on the honey you will harvest. During the winter, on a sunny day with no wind, have a very quick peek under the inner cover to see if you need to replace the patty. Chances are a patty placed on the hive in early November will last the bees over two to three months.

Here's my recipe for grease patties:

1½ pounds of solid vegetable shortening (such as Crisco)

4 pounds of granulated sugar

½ pound honey

⅓ cup of mineral salt (the orange/brown salt available at farm supply stores — it's used to feed to livestock). Pulverize the salt in a blender, breaking it into a fine consistency.

Mix all these ingredients together until smooth. Form into about a dozen hamburger-size patties. Unused patties may be stored in a resealable plastic food bag and kept frozen until ready to use.

NOTE: As an option, you may add 45 milliliters (1.5 ounces) of natural wintergreen oil to the mixture, provided that you're not using this treatment while honey for human consumption is on the hive.

Essential oils

A number of interesting studies have tested the effectiveness of using essential oils as a means of controlling mite populations. Essential oils are those natural extracts derived from aromatic plants such as wintergreen, spearmint, lemon grass, and so on. These oils are available from health-food stores and companies that sell products for making soap.

Pioneering work on the use of essential oils in honey bee hives has been conducted by Bob Noel and Dr. Jim Amrine (West Virginia University). Their experimentation led to discoveries about the use of natural oils in killing off mite infestations in the hive without having any detrimental impact on the bees. A number of different ways exist for using essential oils in the hive.

For up-to-date information and links to thousands of sites devoted to varroa and tracheal mite studies, visit Dr. Amrine's Web site: www.wvu.edu/~agexten/varroa, and Bob Noel's Web site: www.hereintown.net/~rnoel/main.

Honey B Healthy

An all-natural product on the market that's well worth mentioning is Honey B Healthy. It contains pure essential oils (spearmint and lemon grass oils) and is sold as a concentrated food supplement (see figure that follows) that's added to the sugar syrup you feed your bees in the spring and autumn. It was developed by Bob Noel and Jim Amrine, pioneers in the use of essential oils to control mite infestations. The manufacturers make no claims about Honey B Healthy's ability to kill mites, but field tests indicate that this product keeps bees healthy and strong even in the presence of varroa and tracheal mites. I use it religiously, and urge you to give it a try. For more information about the product, visit www.bee-commerce.com.

Formic acid (Apicure)

Formic acid is the stuff that I mentioned earlier in this chapter as a treatment option for varroa mites. It also controls tracheal mites (especially when used in autumn). But formic acid is wickedly caustic, tricky to administer, and may not even be approved for use in your state. I don't suggest that new beekeepers use it.

Wax Moths

Wax moths can do large-scale damage in a weak hive (see the color section, "Spotting Problems," of this book). But they don't usually become a problem in a strong and healthy hive, because bees continually patrol the hive and remove any wax moth larvae they find. If you see wax moths, therefore, you probably have a weak colony. So keeping your bees healthy is the best defense against wax moths.

The story is different when comb is stored for winter. With no bees to protect these combs, the wax is highly susceptible to invasion by wax moths. Steps must be taken to keep the moths from destroying the combs over the winter (see Chapter 12).

Small Hive Beetle

More bad news for bees — besides the Africanized Honey Bees (AHB; see Chapter 9) — came in 1998 when the small hive beetle was discovered in Georgia. Most common beetles that wander in and out of a hive are not a problem, so don't panic if you see some. But the small hive beetle, also originally from Africa, is an exception. The larvae of this beetle eat wax, pollen, honey, bee brood, and eggs. In other words, they gobble up nearly everything in sight. The beetles also — yuck! — defecate in the bees' honey, causing it to ferment and ooze out of the comb. Things can get so nasty that the entire colony may pack up and leave. Who can blame them?

Determining whether you have a small hive beetle problem

First of all, the small hive beetle (so far) is limited to the southeastern United States. It isn't yet certain whether this tropical insect can survive in other colder states. But if you're in Florida, Georgia, South Carolina, or North Carolina, be on the lookout for little black or dark brown beetles scurrying across combs or along the inner cover and bottom board (see Figure 11-11). You may even notice the creamy larvae on the combs and bottom board.

Figure 11-11: The small hive beetle has become a significant problem for beekeepers in some southern states.

Courtesy of Bee Culture Magazine

How to control the small hive beetle

First of all, keeping your colonies strong and healthy is your best natural defense. In addition, you need to destroy any small beetles you see during routine inspections. If infestation levels appear heavy, medicating your hive may be necessary. Presently the only approved treatment for the small hive beetle is *coumaphos* (sold under the brand name of CheckMite+). Because of various government restrictions on the use of this product, it is approved for use only on the small hive beetle in emergency situations (see the warnings about CheckMite+ mentioned earlier in this chapter).

If you suspect that you have the small hive beetle, contact your state apiary inspector. It's important that you do your part to keep this new pest from spreading across the country. The inspector will let you know what kinds of treatments are legal in your state.

Ants, Ants, and More Ants

Ants can be a nuisance to bees. A few ants here and there are normal, and a healthy colony keeps the ant population under control. But every now and then things can get out of hand, particularly when the hive is too young or too weak to control the ant population. Sometimes simply more ants are around than the colony can handle. When ants overrun a colony, the bees will *abscond* (leave the hive). But you can take steps to control the ant population *before* it becomes a crisis. Two things that you can do if you notice more than a few dozen ants in the hive are

✔ **Sending cinnamon to the rescue.** Purchase a large container of ground cinnamon from a restaurant supply company. Sprinkle the cinnamon liberally on the ground around the hive. Sprinkle some on the inner cover. Your hive will smell like a giant breakfast doughnut. Yummy! The bees don't mind, but the ants don't like it and stay away. Remember to reapply the spice after the rain washes it away.

✔ **Creating a moat of oil.** This technique is a great defense against ants. You'll need a hive stand with 18-inch legs. (This is a good idea even if you don't have an ant problem, because raising the hive off the ground is a back-saver for you!). Place each of the stand's four legs in a tin can — old tuna cans are fine. Fill the cans with motor oil. Old or new oil . . . it doesn't matter which you use. The ants won't be able to cross the "moat" of oil and thus are unable to crawl up into the hive (see Figure 11-12).

Figure 11-12:
Placing the legs of your elevated hive stand in cans of motor oil prevents ants from marching into your hive.

Bear Alert!

Do bears like honey? Indeed they do! And they simply crave the sweet honey bee brood. (I've never tried it myself, but I suspect it's sweet.) If bears are active in your area (they're in many states within the continental U.S.), taking steps to protect your hive from these lumbering marauders is a necessity. If they catch a whiff of your hive, they can do spectacular and heartbreaking damage, smashing apart the hive and scattering frames and supers far and wide (see Figure 11-13). What a tragedy to lose your bees in such a violent way. Worse yet, you can be certain that once they've discovered your bees, they'll be back, hoping for a second helping.

The only really effective defense against these huge beasts is installing an electric fence around your apiary. Anything short of this just won't do the trick.

If you're ever unlucky enough to lose your bees to bears, be sure to contact your state or local conservation department. You may qualify for remuneration for the loss of your bees. And the department may provide financial assistance for the installation of an electric fence.

Figure 11-13:
These beehives were shattered to smithereens by a hungry bear.

Raccoons and Skunks

Raccoons are clever animals. They easily figure out how to remove the hive's top to get at the tasty treats inside. Placing a heavy rock on the hive's outer cover is a simple solution to a pesky raccoon problem.

Skunks are insect eaters by nature. When they find insects that have a sweet drop of honey in the center . . . bonanza! Skunks and their families visit the hive at night and scratch at the entrance until bees come out to investigate. When they do . . . they're snatched up by the skunk and . . . gulp! Skunks can put away quite a few bees during an evening's banquet. In time they can decimate your colony. These raids also make your bees decidedly more irritable and difficult to work with. You need to put a quick end to skunk invasions.

Putting your hive on an elevated stand is the best solution for skunk invasions. The skunk then must stand on his hind legs to reach the hive's entrance. That exposes his tender underbelly to the bees — and have no doubts, the bees know what to do next!

You may wonder how the bees feel about skunk scent and whether or not it bothers them. I can't really give you a good answer to this: My bees have never told me how they feel about the smell, nor do I think I've ever known a skunk to spray a hive. But my dog has some stories to tell!

Another solution is hammering a bunch of nails through a plank of plywood (about two-feet square) and placing it in front of the hive with the nail points sticking up. No more skunks. Just be sure *you* remember the plank's there when you go stomping around the hive!

Keeping Out Mrs. Mouse

When the nighttime weather starts turning colder in early autumn, mice start looking for appropriate winter nesting sites. A toasty warm hive is a desirable option. The mouse may briefly visit the hive on a cool night when bees are in a loose cluster. During these exploratory visits the mouse marks the hive with urine so she can find it later on. When winter draws nearer, the mouse returns to the marked hive and builds her nest for the winter.

I can assure you that you don't want this to happen. Mice do extensive damage in a hive during the winter. They don't directly harm the bees, but they destroy comb and foundation and generally make a big mess. They usually leave the hive in early spring, long before the bees break winter cluster and chase them out or sting them to death (see Chapter 2 for the morbid consequences of getting caught). Nesting mice isn't the surprise you want to discover during your early spring inspection. Anticipate mouse problems and take these simple steps to prevent them from taking up winter residence in your hive:

1. As part of winterizing your hive, use a long stick or a wire coat hanger to "sweep" the floor of the bottom board, making sure that no mouse already has taken up residence. Shoo them out if they have.

2. When you're sure your furry friends are *not* at home, secure a metal *mouse guard* along the entrance of the hive (see Figure 11-14). This metal device enables bees to come to and fro and provides ample winter ventilation, but the mouse guard's openings are too small for Mrs. Mouse to slip through.

   Using a wooden entrance reducer as a mouse guard doesn't work. The mouse nibbles away at the wood and makes the opening just big enough to slip through.

Figure 11-14:
Installing a metal mouse guard prevents mice from nesting in your hive during winter.

Some Birds Have a Taste for Bees

If you think you notice birds swooping at your bees and eating them, you may be right. Some birds have a taste for bees and gobble them up as bees fly in and out of the hive. But don't be alarmed. The number of bees that you'll lose to birds probably is modest compared to the hive's total population. No action need be taken. You're just witnessing nature's balancing act.

Part V
Sweet Rewards

The 5th Wave   By Rich Tennant

In this part . . .

These chapters deal with the sweet rewards of bee-keeping! I give you a step-by-step approach for harvesting, bottling, and marketing your honey. I also tell you about other valuable products that you can harvest from your bees.

Chapter 12

Getting Ready for the Golden Harvest

*I*t all comes down to honey. That's why most people keep bees. For eons honey has been highly regarded as a valuable commodity. And why not? No purer food exists in the world. It's easily digestible, a powerful source of energy, and simply delicious. In many countries, honey even is used for its medicinal properties. The honey bee is the only insect that manufactures a food that we eat. And we eat a lot of it — more than one million tons are consumed worldwide each year.

What a thrill it is to bottle your first harvest! You'll swear that you've never had honey that tastes as good as your own. And you're probably right. Commercial honey can't compare to homegrown. Most supermarket honey has been blended, cooked, and ultrafiltered. Yours will be just the way the bees made it, and packed with aroma and flavor. I'm getting hungry just thinking about it.

In this chapter I'll help you plan for the big day — your first honey harvest. You'll need to consider the type of honey you want to produce, the tools you'll need, the amount of preparation you'll have to do, and what you'll need for marketing. So let's get started.

Knowing How Much Honey to Expect

In your first year, don't expect too much of a honey harvest. A newly established colony doesn't have the benefit of a full season of foraging. Nor has it had an opportunity to build its maximum population. I know that's disappointing news. But be patient. Next year will be a bonanza!

Beekeeping is like farming. The actual yield depends upon the weather. Many warm, sunny days with ample rain results in more flowers and greater nectar flows. When gardens flourish, so do bees. If Mother Nature works in your favor, a hive can produce 60 to 100 pounds of surplus honey (that's the honey you can take from the bees), or more. But remember that your bees need you to leave about 60 pounds of honey for their own use during winter.

For a hive to produce that much surplus honey is amazing when you consider that honey bees fly more than 50,000 miles and visit more than 2 million flowers to gather enough nectar to make a pound of honey.

Styling Your Honey: What Do You Want?

The flavor of honey your bees make is likely more up to the bees than you. You certainly can't tell them which flowers to visit. See Chapter 3 for a discussion of where to locate your hive when you want a particular flavor of honey.

Unless you put your hives on a farm with acres of specific flowering plants, your bees will collect myriad nectars from many different flowers, which results in a delicious honey that's a blend of the many flowers in your area. Your honey can be classified as *wildflower honey*. Note also that eating such honey is an effective way to fend off local pollen allergies — a natural way of inoculating yourself. See Chapter 3 for more information on different kinds of honey.

Then you need to decide what *style* of honey you want to package, because that influences some of the equipment that you use.

Deciding on Extracted, Comb, Chunk, or Whipped Honey

What style of honey do you plan to harvest? You have several different options. Each impacts what kind of honey harvesting equipment you purchase, because specific types of honey can be collected only by using specific tools and honey-gathering equipment. If you have more than one hive, you can designate each hive to produce a different style of honey. Now that sounds like fun!

Honey (see Figure 12-1 for the different types) never should be refrigerated, because cold temperatures accelerate crystallization. In time, however, nearly all honeys form granulated crystals, regardless of the temperature. Crystallized honey can be easily liquefied by placing the jar in warm water, or by gently heating in the microwave for a couple of minutes.

Extracted honey

Extracted honey is by far the most popular style of honey consumed in the United States. Wax cappings are sliced off the honeycomb and liquid honey is removed (extracted) from the cells by centrifugal force. The honey is strained and then put in containers (see Figure 12-2). The beekeeper needs an uncapping knife, extractor (spinner), and some kind of sieve to strain out the bits of wax and the occasional sticky bee.

Figure 12-1: Different types of honey.

Courtesy of National Honey Board

*Courtesy of National
Honey Board*

Comb honey

Comb honey is honey just as the bees made it . . . still in the comb (see the photo on the last page of the color insert). Encouraging bees to make this kind of honey is a bit tricky. You need a very strong nectar flow to get the bees going. Watch for many warm sunny days and just the right amount of rain to produce a bounty of flowering plants. But harvesting comb honey is less time consuming than extracted honey. You simply remove the entire honeycomb and package it. You eat the whole thing: the wax and honey. It's all edible! A number of nifty products facilitate the production of comb honey (but more on that later in this chapter).

Chunk honey

Sometimes called *cut comb, chunk honey* refers to chunks of honeycomb that are placed in a wide-mouthed bottle and then filled with extracted liquid honey.

Whipped honey

Also called *creamed honey, cremed honey, spun honey, churned honey, candied honey,* or *honey fondant, whipped honey* is a semisolid style of honey that's popular in Europe. In time, all honey naturally forms coarse granules or crystals. By controlling the crystallization process you can produce fine crystals and create a smooth, spreadable product.

Granulated honey is honey that has formed sugar crystals. You make *whipped* honey by blending nine parts of extracted liquid honey with one part of finely granulated (crystallized) honey. The resulting consistency of whipped honey is thick, ultrasmooth, and can be spread on toast like butter. Making it takes a fair amount of work, but it's worth it!

Selecting the Right Equipment for the Job

Once you decide what style of honey you want your bees to make (extracted, comb, chunk, or creamed) you need to get hold of the appropriate kind of equipment. This section discusses the various types that you'll need depending upon the style of honey you want to harvest.

How to make whipped honey: The Dyce Method

The Dyce Method is a process used to control the crystallization of honey. It was developed and patented by Elton J. Dyce in 1935. The process (described here) results in a nice, smooth whipped honey:

✔ Heat honey to 120 degrees F. This kills yeast cells that always are present in honey. Yeast causes fermentation, and its presence can inhibit a successful result when making whipped honey). Stir the honey gently and constantly to avoid overheating. Be careful not to introduce air bubbles.

✔ Using a two-fold thickness of cheesecloth as a strainer, strain honey to remove foreign material and wax.

✔ Heat honey again, this time to 150 degrees F. Don't forget to stir continuously.

✔ Strain honey another time to remove all visible particles. Again, you can use a two-fold thickness of cheesecloth as a strainer.

✔ Cool honey as rapidly as possible. You can place honey in a container and "float" it in an ice water bath to speed the cooling process. Stir gently as honey cools. Continue cooling until the temperature of the honey reaches 75 degrees F.

✔ Add some finely crystallized honey to promote a controlled crystallization of your whipped honey — it's kind of like adding a special *yeast culture* when making sourdough bread. Introduce these seed *crystals* by adding 10 percent (by weight) of processed granulated honey. *Granulated honey* is processed by breaking down any coarse crystals into finely granulated crystals. This can be accomplished by *fracturing* the crystallized honey in a meat grinder or a food processor.

✔ Place mixture in a cool room (57degrees F). Complete crystallization occurs in about a week.

✔ After a week, run mixture through the grinder (or food processor) one more time to break up any newly formed crystals.

✔ Bottle and store in a cool dry room.

(Information courtesy of National Honey Board)

Honey extractors

Essentially, an extractor is a device that spins honey from the comb using centrifugal force (see Figure 12-3). Extractors come in different sizes and styles to meet virtually every need and budget. Hand-crank models or ones with electric motors are available. Small ones for the hobbyist with a few hives, huge ones for the bee baron with many hives, and everything in between can be found. Budget extractors are made entirely of plastic, and rugged ones are fabricated from food-grade stainless steel. Keep in mind, however, that a good quality stainless steel extractor will far outlast a cheap one made of plastic. So get the best one that your budget allows. Look for a model that accommodates at least four frames at a time. Backyard beekeepers can expect to pay $250 to $395 for a nice quality extractor.

You may not have to buy an extractor. Some local beekeepers, beekeeping clubs, and nature centers rent out extractors. So be sure to call around and see what options you have. Ultimately, you may want to invest in your own. My advice: If you're able to, rent or borrow an extractor during your first season. From the experience you gain, you'll be better able to choose the model and style of extractor that best meets your needs.

Figure 12-3: This hand-crank, stainless steel extractor extracts up to six shallow frames at a time.

Uncapping knife

The wax cappings on the honeycomb form an airtight seal on the cells containing honey — like a lid on a jar. Before honey can be extracted, the "lids" must be removed. The easiest way is by using an uncapping knife. These electrically heated knives slice quickly and cleanly through the cappings (see Figure 12-4).

Alternatively, you can use a large serrated bread knife. Heat it by dipping in hot water (be sure to wipe the knife dry before you use it to prevent any water from getting into your honey).

Honey strainer

The extracted honey needs to be strained before you bottle it. This step removes the little bits of wax, wood, and the occasional sticky bee. Any kind of conventional kitchen strainer or fine-sieved colander will suffice. Nice stainless steel honey strainers (see Figure 12-5) are made just for this purpose and are available from your beekeeping supplier.

Or you can use a disposable paint strainer (available at you local paint supply store). They do the trick just fine, and fit nicely over a 5 gallon plastic bucket.

Figure 12-4:
An electrically heated uncapping knife makes short order of slicing wax cappings off honeycomb.

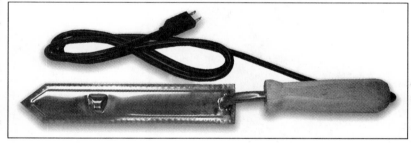

Figure 12-5:
A double stainless steel honey strainer (like this one) is an effective way to clean up your honey before bottling it.

Other handy gadgets for extracting honey

Here are a few of the optional items that are available for extracting honey. None are essential, but all are useful niceties.

Double uncapping tank

The double uncapping tank is a nifty device that is used to collect the wax cappings as you slice them off the comb. The upper tank captures the cappings (this wax eventually can be rendered into candles, furniture polish, cosmetics, and so on). The tank below is separated by a wire rack, and collects the honey that slowly drips off the cappings. Some say the sweetest honey comes from the cappings! The model shown in Figure 12-6 also has a honey valve in the lower tank.

Figure 12-6:
A double uncapping tank helps you harvest wax cappings. It reclaims the honey that drains from the cappings.

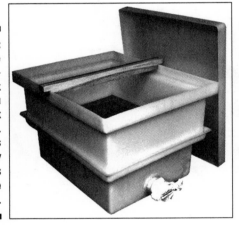

Uncapping fork

An uncapping fork is used to scratch open cappings on the honeycomb (see Figure 12-7). It can be used in place of or as a supplement to an uncapping knife (the fork opens stubborn cells missed by the knife).

Bottling bucket

Five-gallon bottling buckets are made with food-grade plastic and include a honey gate. They come with airtight lids and are handy for storing and bottling honey. Each pail holds nearly 60 pounds of honey. I always keep a few of them on hand (see Figure 12-8).

Figure 12-7:
An uncapping fork is a useful tool for opening cappings missed by your uncapping knife.

Figure 12-8:
The honey gate valve on this 5-gallon bucket makes bottling your honey a breeze.

Solar wax melter

Aside from honey, one of the most important products of the hive is beeswax. From the wax you can make candles, furniture polish, and cosmetics. Your primary harvest of wax is the result of the cappings that you cut from the comb during the honey extraction process. These cappings (and any burr comb that you trim from the hive during the year) can be placed in a solar wax melter and melted. A single hive yields enough surplus wax to make a few candles and some other wax products (such as wax polish or hand cream).

You can obtain a *solar wax melter* by purchasing it from your bee supplier or by making one yourself. It typically consists of a wooden box containing a metal pan, covered with a glass lid. The sun melts the wax, which is collected in a tray at the base of the unit. It's a handy piece of equipment if you plan to make use of all that wax.

Comb honey equipment

Harvesting comb honey boils down to two basic equipment choices, using section comb cartridges or the cut-comb method. Either works fine. You'll need special equipment on your hives to produce these special kinds of honey. See Chapter 4 for additional information about comb honey production.

Section comb cartridges

Honeycomb kits consist of special supers containing wooden or plastic section comb cartridges. Each cartridge contains an ultrathin sheet of wax foundation. Using them enables the bees to store honey in the package that ultimately is used to market the honeycomb. My favorite kit — Ross Rounds — makes circular section comb in clear plastic containers. This is a product with enormous eye-appeal!

You typically need a strong nectar flow to encourage the bees to make any kind of section comb honey.

Cut comb

The cut-comb method uses conventional honey supers and frames, but it also uses a special foundation that is ultrathin and unwired. Once bees fill the frames with capped honey, the comb is cut from the frames. You can use a knife, or a *comb honey cutter,* which looks like a square cookie cutter and makes the job easy.

Honey containers

Select an attractive package for your honey (jar, bottle, and so on). Many options are available to you, and quite frankly, any kind of container will do. Clear containers are best, because customers want to be able to see what they're getting. Either plastic or glass is okay to use. You can purchase all kinds of specialized honey bottles from your beekeeping supplier. Or simply use the old mayonnaise and jam jars that you've been hoarding.

Planning Your Honey Harvest Setup

Giving some thought to where you plan to extract and bottle your honey is important. You can use your basement, garage, toolshed, or even your kitchen. You don't need a big area. If you have only a few hives, harvesting is a one-person job. But be prepared — you'll likely get plenty of volunteers who want to help out. The kids in my neighborhood are eager to lend a hand in exchange for a taste of my *liquid candy.* The guidelines in the following list will help you choose the best location:

✓ The space you choose must be absolutely beetight. That is to say, you don't want any bees getting into the space where you're working. The smell of all that honey will attract them, and the last thing you want is hundreds (or thousands) of ravenous bees flying all about.

✓ Never, ever attempt to harvest your honey outdoors. If you do, disaster is imminent! In short order you'll be engulfed by thousands of bees, drawn by the honey's sweet smell.

✓ Set up everything in advance, and arrange your equipment in a way that complements the sequential order of the extraction process (see Figure 12-9).

✓ Have a bucket of warm water — better yet, hot and cold running water — and a towel at the ready. Life gets sticky when you're harvesting honey, and the water is a welcome means for rinsing off your hands and uncapping knife.

✓ If you're using an electric uncapping knife, you'll need an electrical outlet. But remember that water and electricity don't mix well, so be careful!

✓ Place newspaper on the floor. This little step saves time during cleanup. If your floor is washable, that really makes life easy!

Figure 12-9: Here's the setup, left to right, in my garage for extracting honey.

Branding and Selling Your Honey

Before you harvest your first crop of honey, you may want to give some advanced thought to the label you will put on it. You may even want to sell your honey. After all, a hundred or more bottles of honey may accommodate more toast than your family can eat! The following sections describe some ideas to help you think this through.

Creating an attractive label

An attractive label can greatly enhance the appearance and salability of your honey. It also includes important information about the type of honey and who packages it (you!). Generic labels are available from your beekeeping supplier. Or you can make your own custom label. I easily reproduce my labels (see Figure 12-10) using my computer's printer and an appropriate size of blank, self-adhesive labels.

Figure 12-10:
Here's my honey label — simple and to the point.

You must include a few important bits of information on your label (assuming that you're planning to sell your honey). Listed below are label requirements that you need to keep in mind (see Figure 12-11). You must

- ✔ State what the container contains: HONEY.
- ✔ Include your name and address (as the producer).
- ✔ Report the net weight on the lower 20 percent of the label using a dual weight declaration, for example, NET WEIGHT 16 OZ. (1 lb). Federal law mandates it.

In addition to these requirements, I recommend

  ✔ Adding information about the type of honey in the package (for example, WILDFLOWER) and some marketing propaganda about the pure and wholesome nature of the product.

  ✔ Including information about the nutritional value (not usually required by law). I think it makes the product far more professional looking. For more on the wording and design of a nutritional information label, see Figure 12-11.

Go to your local market and make mental notes about commercial honey labels. Which ones appeal to you? What about them makes them look so attractive? What kind of image or graphic is used? Which colors look best? Borrow ideas shamelessly from the ones you like best — but be careful not to steal anything that may be trademarked!

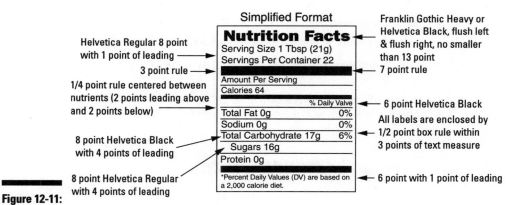

Figure 12-11: This is the standard layout and nutritional information that can appear on your label. Information in this example is based on a one-pound jar of honey.

More detailed information about creating a distinctive label is available from the National Honey Board, 390 Lashley St., Longmont, CO 80501-6045. Its Web address is: www.nhb.org.

Finding places to market your honey

An independently owned food market in your neighborhood may be interested in selling your honey. Honey is a pure and natural food, and you don't need a license to package and sell it (more detailed information is available from the National Honey Board, 390 Lashley St., Longmont, CO 80501-6045; see Web address in the previous section. Here are some other ideas:

- ✔ Check out health food stores. They're always looking for a source of fresh, local honey.
- ✔ Gift stores, craft shops, and boutiques are good places to sell local honey.
- ✔ Put up an attractive sign in front of your house: HONEY FOR SALE.
- ✔ Sell your honey at the local Farmers' Market.
- ✔ Don't forget to consider church fairs, synagogue bazaars, and gardening centers.
- ✔ And by all means GIVE a bottle to all your immediate neighbors. It's the right thing to do, and a great public relations gesture.

Selling your honey on the Web

This is the web generation. So why not set up a simple Web site to sell your honey all over the world? That's what I do! Remember that plastic honey jars are lighter to ship and less fragile than glass. This may be the time to invest in *Creating Web Pages For Dummies* for more information on that particular subject.

Chapter 13

Honey Harvest Day

*T*he day you've anticipated all year is finally here and it's time to reap the rewards of all your efforts (actually the bees did most of the work, but go ahead and take the credit anyway). It's time for the honey harvest!

And what better excuse could you ever have for having a party! Harvest day always is a big event at my house — a day when neighbors and friends gather to lend a hand and get a free sample from the magic candy machine. I schedule an open house on honey harvest day, inviting anyone interested to stop by, lend a hand, see how it's done, and go home with a bottle of honey still warm from the hive. I have plenty of honey-related refreshments on hand: honey-ginger snaps, and honey-sweetened ice tea and lemonade.

I am assuming that you've decided to harvest extracted honey. That's what I recommend for new beekeepers. Harvesting extracted honey is easier for you and the bees, and you're far more likely to get a substantial crop than if you were trying to produce comb honey. *Comb honey* requires picture-perfect conditions to realize a successful harvest (large hive population, hard working and productive bees, ideal weather conditions, and excellent nectar flows).

Extracted honey refers to honey that is removed from wax comb by centrifugal force (using a device call an extractor). The honey is bottled in a liquid form, as opposed to harvesting comb or chunk honey where the honey is not removed from the wax comb before it's packaged.

Be sure to allow yourself enough time. I set aside an entire weekend for my harvest activities, one afternoon to get the honey supers off the hive and the following day to actually extract and bottle the honey.

Knowing When to Harvest

Generally speaking, beekeepers harvest their honey at the conclusion of a substantial nectar flow and when the hive is filled with cured and *capped* honey (see Figure 13-1). Conditions and circumstances vary greatly across the country. Early one spring, I had an unusually large flow of nectar from a large honey locust tree. My bees filled their honey supers before June. I harvested this rare and delicate white honey in late May. I put the supers back on and got another harvest in the late summer. More typically, I will wait until late summer to harvest my crop (usually mid-September). Where I live (in the northeastern United States) the last major nectar flow (from the asters) is over by September. First-year beekeepers are lucky if get a small harvest of honey by late summer. That's because a new colony needs a year to build up a large enough population to gather a surplus of honey.

I suggest that you take a peek under the hive cover every couple of weeks during summer. Note what kind of progress your bees are making and find out how many of the frames are filled with capped honey.

When a shallow frame contains 80 percent or more of sealed, *capped* honey, you're welcome to remove and harvest this frame. Or, you can practice patience — leave your frames on and wait until one of the following is true:

- ✔ The bees have filled all the frames with capped honey.
- ✔ The last major nectar flow of the season is complete.

Honey in *open* cells (not capped with wax) can be extracted if it is cured. To see if it's cured, turn the frame with the cells facing the ground. Give the frame a gentle shake. If honey leaks from the cells, it isn't cured and shouldn't be extracted. This stuff is not even honey. It's nectar that hasn't been cured. The water content is too high for it to be considered honey. Attempting to bottle the nectar results in watery syrup that is likely to ferment and spoil.

Figure 13-1:
Here's a
beautiful
frame of
capped
honey ready
to be
harvested.

For goodness sakes, don't procrastinate!

You want to wait until the bees have gathered all the honey they can, so be patient. That's a virtue. However, don't leave the honey supers on the hive *too* long! I know, I know! Things tend to get busy around Labor Day. Besides spending a weekend harvesting your honey, you probably have plenty of other things to do. But don't put off what must be done. If you wait too long, one of the following two undesirable situations can occur:

- ✔ After the last major nectar flow and winter looms on the distant horizon, bees begin consuming the honey they've made. If you leave supers on the hive long enough, the bees will eat much of the honey you'd hoped to harvest. Get those supers off the hive before that happens!

- ✔ If you wait *too* long to remove your supers, the weather turns too cold to harvest your honey. In cool weather, honey can thicken and granulate, which makes it impossible to extract from the comb. I discuss this later in this chapter in the "Two common honey-extraction questions" sidebar. Remember that honey is easiest to harvest when it still holds the warmth of summer.

Getting the Bees Out of the Honey Supers

Regardless what style of honey you decide to harvest, you must remove the bees from the honey supers before you can extract or remove the honey. You've heard the old adage, "Too many cooks spoil the broth!" Well, you certainly don't need to bring several thousand bees into your kitchen!

You must leave the bees 60 to 70 pounds of honey for their own use during winter months (less in those climates that don't experience cold winters). But anything they collect more than that is yours for the taking.

To estimate how many pounds of honey are in your hive, figure that each deep frame of capped honey weighs about 7 pounds. If you have ten deep frames of capped honey, you have 70 pounds!

Removing bees from honey supers can be accomplished in many different ways. This section discusses a few of the more popular methods that beekeepers employ. Before attempting any of these methods, be sure to smoke your bees the way you normally would when opening the hive for inspection. (See Chapter 6 for information on how to use your smoker properly.)

The bees are protective of their honey during this season. Besides donning your veil, now's the time to wear your gloves. If you have somebody helping you, be sure they are also adequately protected.

Shakin' 'em out

This bee-removal method involves removing frames (one by one) from honey supers and then shaking the bees off in front of the hive's entrance. The cleared frames are put into an empty super. Be sure that you cover the super with a towel or board to prevent bees from robbing you of honey. Alternatively, you can use a bee brush (see Chapter 4) to gently brush bees off the frames.

Note that the cells on comb tend to slant downward slightly — to better hold liquid nectar. Therefore, when brushing bees, you should always brush bees gently *upward* (never downward). This little tip helps prevent you from injuring or killing bees that are partly in a cell when you're brushing.

Shaking and brushing bees off frames isn't the best option for the new beekeeper, because it can be quite time consuming, particularly when you have a lot of supers to clear. Besides the action can get pretty intense around the hive during this procedure. The bees are desperate to get back into those honey frames, and, because of their frenzy, you can become engulfed in a fury of bees. Don't worry — just continue to do your thing. The bees can't really hurt you, provided you're wearing protective gear.

Blowin' 'em out

One fast way to remove bees from supers is by blowing them out, but they don't like it much. Honey supers are removed from the hive (bees and all) and stood on end. By placing them 15 to 20 feet away from the hive's entrance and using a special bee blower (or a conventional leaf blower), the bees are blasted from the frames at 200 miles an hour. Although it works, to be sure, the bees wind up disoriented and *very* irritated. Oh goodie. Again, I wouldn't recommend this method for the novice beekeeper.

A bee blower is basically the same as a conventional leaf blower, just packaged differently and usually more expensive.

Using a bee escape board

Yet another bee-removal method places a bee escape board between the upper deep-hive body and the honey supers that you want to clear the bees from. Various models of *escape boards* are available, and all work on the same principle: The bees can travel down to the brood nest, but they can't immediately figure out how to travel back up into the honey supers. It's a one-way trip. (See Figure 13-2 for an example of a triangle bee escape with a maze that prevents the bees from finding their way back up into the honey supers.)

Bee escapes work okay, but it takes a few days for the bees to be cleared from the honey supers. You must install the escape boards 48 hours *before* you plan to remove the honey supers. It takes that long to clear the bees. And, the thing is, you can't leave the escape board on for *more* than 48 hours, or the bees eventually solve the puzzle and find their way back into the supers. For me, a weekend beekeeper, the timing of all this is quite impractical.

Figure 13-2:
This triangle bee escape enables bees to easily travel down into the top deep, but it takes the bees a while to figure out how to get back up into the honey supers.

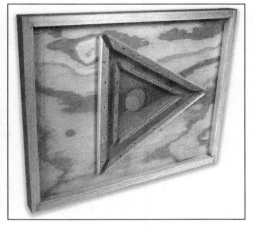

Fume board and bee repellent

Here's my favorite method — a fume board and bee repellent! It's a fast and highly effective. And it's made even more desirable because of a wonderful new product on the market (more about that later in this section).

A *fume board* looks like an outer cover with a flannel lining. A liquid bee repellent is applied to the flannel lining and the fume board is placed on top of the honey supers (in place of the inner and outer covers). Within 5 minutes, the bees are repelled out of the honey supers and down into the brood chamber. Instant success! The honey supers can then be safely removed and taken to your harvesting area.

In the past, chemicals used as repellents (either butyric propionic anhydride or benzaldehyde) have been hazardous in nature. They're toxic, combustible, and may cause respiratory damage, central nervous system depression, dermatitis, and liver damage. Need I say more? It's simply nasty stuff to have around the house.

In addition, the stench of each of these products is more than words can politely express. All that recently changed with the introduction of a product called Fischer's Bee-Quick (see Figure 13-3). This product repels the bees, but it is nonhazardous and made entirely from natural ingredients. Best of all, its almond-vanilla scent smells good enough to be a dessert topping!

Figure 13-3: A safe and fast way to get bees out of honey supers is to use a fume board with Fischer's Bee Quick.

Here are step-by-step instructions for using a fume board with Fischer's Bee-Quick:

1. **Smoke your hive as you would for a normal inspection.**

2. **Remove the outer and inner covers, and the queen excluder.**

3. **Use your smoker on the top honey super to drive the bees downward.**

4. **Sprinkle Fischer's Bee Quick on the fume board's felt pad in a zigzag pattern (as if corresponding to spaces between the frames) across the full width of the fume board.**

   Don't overdo it. About one ounce or less should do the trick (use more in cold, cloudy weather, less in hot, sunny weather). When in doubt . . . use less.

5. **Place fume board on the uppermost honey super and wait 3 to 5 minutes for the bees to be driven out.**

6. **Remove the top honey super. Put the super aside and cover it with a towel or extra hive cover to avoid robbing (discussed in the next section).**

7. **Repeat the process for subsequent honey supers.**

Keep in mind that a shallow super full of capped honey can weigh 30 to 40 pounds. You'll have a heavy load to move from the beeyard to wherever you'll be extracting honey. So be sure to save your back and take a wheelbarrow or hand truck with you when removing honey supers from the hive.

Honey Extraction 101

Once the bees are out of the honey supers, you need to be prepared to process your honey as soon as possible (within a few days). Doing so minimizes the chance of a wax moth infestation, discussed in the "Controlling wax moths" section later in this chapter. Besides, extracting honey is easier to do when the honey is still warm from the hive.

For a description of the various tools used in the honey-extraction process (uncapping knife, honey extractor, bucket of warm water, a towel, and so forth) see Chapter 12.

Follow this procedure when extracting honey from your frames:

1. **One by one, remove each frame of capped honey from the super.**

   Hold the frame vertically over the uncapping tank and tip it slightly forward. This helps the cappings fall away from the comb as you slice them.

2. **Use your electric uncapping knife to remove the wax cappings and expose the cells of honey.**

   A gentle side-to-side slicing motion works best, like slicing bread. Start a quarter of the way from the bottom of the comb, slicing upward (see Figure 13-4). Keep your fingers out of harm's way in the event the knife slips. Complete the job with a downward thrust of the knife to uncap the cells on the lower 25 percent of the frame.

3. **Use an uncapping fork (also called *cappings scratcher*) to get any cells missed by the knife.**

   Flip the frame over, and use the same technique to do the opposite side.

   I discuss what you should do with the wax cappings, particularly if you want to use them for craft purposes, later in this chapter.

4. **When the frame is uncapped, place it vertically in your extractor (see Figure 13-5).**

   An *extractor* is a device that spins the honey from the cells and into a holding tank.

   Once you've uncapped enough frames to fill your extractor, put the lid on and start cranking. Start spinning slowly at first, building some speed as you progress. Don't spin the frames as fast as you can, because extreme centrifugal force may damage the delicate wax comb. After spinning for five to six minutes, turn all the frames to expose the opposite sides to the outer wall of the extractor. After another five to six minutes of spinning, the comb will be empty. The frames can be returned to the shallow super.

5. **As the extractor fills with honey, it becomes increasingly difficult to turn the crank (the rising level of honey prevents the frames from spinning freely), so you need to drain off some of the harvest.**

   Open the valve at the bottom of the extractor and allow the honey to filter through a honey strainer and into your bottling bucket.

6. **Use the valve in the bottling bucket to fill the jars you've designed for your honey.**

   Brand it with your label and you're done! Time to clean up.

Figure 13-4:
Removing wax cappings with an electric uncapping knife.

Honey is *hygroscopic* — meaning that it absorbs moisture from the air. On the positive side, this is why baked goods made with honey stay moist and fresh. On the negative side, this means you must keep your honey containers tightly sealed, otherwise your honey will ferment.

Figure 13-5:
Place the
uncapped
frame
vertically
in the
extractor.

Having enough honey jars and lids on hand is important. Standard honey jars are available in 1-, 2- and 5-pound sizes. You can estimate that you'll harvest about 30 pounds of honey from each shallow honey super (assuming all of the frames are full of honey).

You can store your frames of honey (briefly) before extracting

Extracting your honey as soon as you can after taking the honey supers off the hive is best. I try to schedule my extraction activities either on the same day that I remove the supers from the hive or the day following removal. If that isn't practical, you can temporarily store your shallow supers in a bee-tight room where the temperature is between 80 and 90 degrees F. Honey that's kept warm is much easier to extract and strain and is far less likely to granulate. A new beekeeper with one or two hives can expect to spend 3 to 5 hours on the extraction process.

Two common honey-extraction questions

Here's a question I'm frequently asked: *When I spin frames in my honey extractor, the entire unit wobbles uncontrollably, dancing across the floor. How do I prevent this from happening?*

This shimmy happens when the load in the extractor is unbalanced. Make sure you have a frame in each of the basket's slots. Try redistributing the weight in the basket. It's kind of like rearranging the wash load in the washing machine during the spin cycle. Some extractors can be bolted to a sturdy table — that helps.

Another frequently asked question I receive is: *The honey in some of my shallow frames has granulated. I cannot remove the granulated honey using my extractor. How can I extract granulated honey from the honeycomb?*

Unfortunately, no practical way to extract granulated honey from the honeycomb exists without destroying the honeycomb (melting the comb and wax, and then separating the wax once it floats to the top and solidifies). I suggest that you put supers with granulated honey back on the hives in the early spring. Before you do, scratch the cappings with an uncapping fork, exposing the honey to the bees. They'll consume the granulated honey and leave the combs sparkling clean and ready for a new harvest. **Note:** Don't feed your bees granulated honey in late autumn, because it can give them dysentery.

Cleaning Up After Extracting

Never store extracted frames while they're wet with honey. You'll wind up with moldy frames that have to be destroyed and replaced next year. You've got to clean up the sticky residue on the extracted frames. How? Let the bees do it!

At dusk, place the supers with the empty frames on top of your hive (sandwiched between the top deep and the inner and outer covers). Leave the supers on the hive for a few days, and then remove them. The bees lick up every last drop of honey, making the frames bone dry and ready to store until next honey season. Be sure to treat your frames with wax moth control before storing them away for the winter (see the next section).

Controlling wax moths

There's a good chance that honey supers stored over the winter will become infested with wax moths. It happens. The adult moth lays eggs on beeswax comb before the hives are stored for the winter. The developing larvae tunnel their way through the wax comb, leaving a crisscrossed pattern of silky trails. In time, all the frames are destroyed and become useless. What a disheartening sight to discover this damage (see the color section of this book for a photo of what wax moths can do to a frame)! You can help prevent wax moth damage in stored honey supers by:

✔ **Fumigating them with PDB.**

My favorite solution is to treat stored supers with wax moth control crystals (paradichlorobenzene, or PDB; see Figure 13-6). This product is available from your bee supplier. Place a tablespoon of PDB crystals on an index card and put the card on the top bars of the super to be stored. Do this for every super that you plan to store, stacking them one on top of another, like so many floors of a building.

Top the whole stack with an outer cover. Make sure there are no gaps or cracks. You want the slowly evaporating crystals to fumigate all the frames while they're stored over the winter. If you've drilled any ventilation holes in the supers, make sure they're taped shut.

Store supers in any area where the air temperature is above 60 degrees F. When spring arrives, and you're ready to place the honey supers on your hive, be sure to air them out for 48 hours before using them.

✔ **Freezing the combs.**

You can destroy wax moth larvae by placing the frames in the deep-freezer for 24 hours. This assumes, of course, that you have a really big freezer! Then put the frames back in the supers and store them in tightly sealed plastic garbage bags. The colder the storage area, the better.

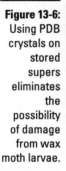

Figure 13-6:
Using PDB crystals on stored supers eliminates the possibility of damage from wax moth larvae.

Harvesting wax

When you extract honey, the cappings that you slice off represent your major wax harvest for the year. You'll probably get one or two pounds of wax for every 100 pounds of honey that you harvest. This wax can be cleaned and melted down for all kinds of uses (See the section on beeswax in Chapter 14). Pound for pound, wax is worth more than honey, so it's definitely worth a bit of effort to reclaim this prize. Here are some guidelines:

1. **Allow gravity to drain as much honey from the cappings as possible.**

   Let the cappings drain for a few days. Using a double uncapping tank greatly simplifies this process.

2. **Place the drained cappings in a 5-gallon plastic pail and top them off with warm (not hot) water.**

   Using a paddle — or your hands — slosh the cappings around in the water to wash off any remaining honey. Drain the cappings through a colander or a honey strainer and repeat this washing process until the water runs clear.

3. **Place the washed cappings in a double boiler and melt the wax.**

   Always use a double boiler for melting beeswax (never melt beeswax directly on an open flame, because it is highly flammable). And *never, ever* leave the melting wax — even for a moment. If you need to go to the bathroom, turn off the stove!

4. **Strain the melted beeswax through a couple of layers of cheesecloth to remove any debris.**

   Remelt and restrain as necessary to remove all impurities from the wax.

5. **The rendered wax can be poured into a block mold for later use.**

   I use an old cardboard milk carton. Once the melted wax has solidified in the carton, it can easily be removed by tearing away the carton. You're left with a hefty block of pure, light-golden beeswax.

For ideas about what you can make with this lovely wax, see Chapter 14.

Part VI
The Part of Tens

"Remind me the next time we're harvesting honey not to sit down and pet the cat afterwards."

In this part . . .

No *For Dummies* book is complete without the Part of Tens. In this part, I offer a collection of fun lists, frequently asked questions, and honey recipes. Not a bad way to squeeze a whole bunch of extra, helpful information into a book!

Chapter 14

Ten Fun Things to Do with Bees

*O*ne of the glorious things about keeping bees is that your interests can expand way beyond the business with your smoker and hive tool. Beekeeping opens up entire new worlds of related hobbies and activities — horticulture, carpentry, biology, and crafts just to name a few. That's been a good thing for me, because living in Connecticut, as I do, the winters were unbearably long when I couldn't play with my bees. I really missed them! But now, having gotten drawn into some of these "related" hobbies, I can hardly find time to sit and think. Here are a few of the bee-related activities whose sirens have beckoned to me over the years.

Making Two Hives From One

If you're like most beekeepers I know, it's only a matter of time before you start to ask yourself, "Gee, wouldn't it be twice as much fun to have twice as many hives?" Well, actually it is. And the neat thing is that you can create a second colony from your existing colony. You don't even have to order another package of bees! Free bees! Ah, but here's the dilemma! You'll need a new queen for your new colony. Strictly speaking, you don't have to order a new queen. You can let the bees make their own; however, ordering a new queen is simply faster and more foolproof. I discuss the nuances of ordering a new queen later in this chapter.

To do this, you first need a strong, healthy hive. That's just what you hope your hive will be like at the start of its second season — boiling with lots and lots of busy bees. The procedure is known as *dividing* or *making a divide*.

Dividing not only enables you to start a new colony, it's also considered good bee management — dividing thins out a strong colony and prevents that colony from swarming.

The best time to make a divide is in the early spring about a month before the first major nectar flow. Follow these steps in the order they are given:

1. **Check your existing colony (colonies) to determine whether you have one that's strong enough to divide.**

    Look for lots of bees, and lots of capped brood (six frames of capped brood and/or larvae is ideal). The situation should look crowded.

2. **Order a new hive setup from your bee supplier.**

    You'll want hive bodies, frames, foundation — the works. You need the elements to build a new home for your new family.

3. **Order a new queen from your bee supplier.**

    Alternatively, you can allow the new colony to raise its own queen (see the sidebar, "Let them make a queen," later in this chapter).

    Your new queen doesn't have to be marked, but having a marked queen is a plus, particularly when you're looking for her because the mark makes her easier to identify. I advise you, a new beekeeper, to let your bee vendor mark your queen. A novice can end up killing a queen by mishandling her.

4. **Put your new hive equipment where you plan to locate your new family of bees.**

    You'll need only to put out one deep-hive body at this point — just like when you started your first colony (see Chapter 5). Remove four of the ten foundation frames and set them aside. You'll need them later.

5. **When your new marked queen arrives, it's time to divide!**

    Smoke and open your existing colony as usual.

6. **Find the frame with the queen and set it aside in a safe place.**

    An extra empty hive body and cover will do just fine.

7. **Now remove three frames of capped brood (frames with cells of developing pupae) plus all the bees that are on each of them.**

    Place these three brood frames and bees in the center of the new hive. I know, I know — that still leaves one slot open because your removed *four* frames of foundation. The extra slot, however, provides the space that you'll need to hang the new queen cage (see Step 8).

8. **Using two frame nails, fashion a hanging bracket for the new queen cage (candy side up) and hang the cage between brood frames in the middle of the new hive.**

Make sure you have removed the cork stopper or metal disc, revealing the candy plug. This is the same queen introduction technique that you used when you installed your first package of bees (see Chapter 5 and the color-photo section of this book).

9. **Put a hive-top feeder on your new colony and fill it with sugar syrup.**

10. **Turn your attention back to the original hive.**

    Carefully put the frame containing the queen back into the colony. Add three of the new foundation frames (to replace the three brood frames that you removed earlier).

11. **Add a hive-top feeder to your original hive and fill it with sugar syrup.**

Congratulations, you're the proud parent of a new colony! But wait, you say, "I've got one new frame of foundation left over." Good. That's what you'll use next week to replace what will then be an empty queen cage.

Making One Hive From Two

Keep in mind that it's better to go into the winter with strong colonies — they have a far better chance of making it through the stressful cold months than do weak ones.

If you have a weak hive, you can combine it with a stronger colony. If you have two weak hives, you can combine them to create a robust colony. But you can't just dump the bees from one hive into another. If you do, all hell will break loose. Two colonies must be combined slowly and systematically so that the hive odors merge gradually — little by little. This is best done late in the summer or early in the autumn (it isn't a good idea to merge two colonies in the middle of the active swarming season).

My favorite method for merging two colonies is the so-called *newspaper method.* A single sheet of newspaper separates the two hives that you'll combine. Follow these steps in the order they are given:

1. **Identify the stronger of the two colonies.**

   Which colony has the largest population of bees? Its hive should become the home of the combined colonies. The stronger colony stays put right where it's now located.

2. **Smoke and open the weaker colony (see Chapter 6 for instructions).**

   Manipulate the frames so that you wind up with a single deep-hive body containing ten frames of bees, brood, and honey. In other words, consolidate the bees and the ten best frames into one single deep.

3. **Smoke and open the stronger hive.**

Remove the outer and inner covers and put a single sheet of newspaper on the top bars. Poke a few holes in the newspaper with a small nail. This helps hive odors pass back and forth between the strong colony and the weak one that you're about to place on top.

4. **Take the hive body from the weak colony (it now contains ten consolidated frames of bees and brood) and place it directly on top of the stronger colony's hive.**

   The perforated sheet of newspaper separates the two colonies (see Figure 14-1).

5. **Add a hive-top feeder and fill it with sugar syrup.**

   The outer cover goes on top.

6. **Check the hive in a week.**

   The newspaper will have been chewed away and the two colonies will have happily joined into one whacking strong colony. The weaker queen is now history and only the stronger queen remains.

7. **Now you have the task of consolidating the three deeps back into two.**

   Go through all the frames, selecting the 20 best frames of honey, pollen, and brood. Arrange these in the lower two deeps. Shake the bees off the ten surplus frames and into the lower two deeps (save these frames and the third hive body as spares).

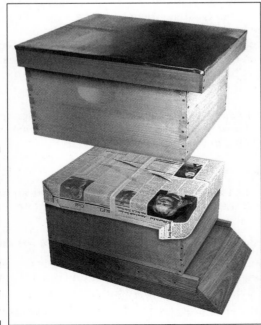

Figure 14-1:
A weak colony can be combined with a strong colony by using the newspaper method.

Let them make a queen

You don't necessarily have to order a queen from a supplier. You can let a colony raise a new queen. But for this to happen, eggs must be present. Make certain some of the frames that you place in a new (queenless) colony include eggs.

The new colony will build queen cells and raise a new queen from these eggs. The cutaway image below shows two developing queen bee larvae in their cells, nestled in creamy white royal jelly.

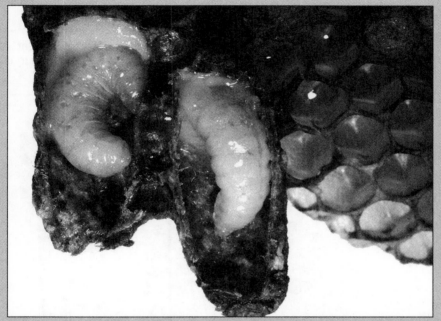

Courtesy of Stephen McDaniel

Be aware that the process of allowing the hive to raise their own queen will delay the productivity of your hive by nearly a month (it takes 16 days to go from egg to adult queen, and another week for her to mate and begin laying eggs). Another thing to consider is that the genetics of your new colony can be uncertain (you have no control over with whom your new virgin queen will mate). The resulting offspring may not be as docile or productive as you hoped. My advice to the *new* beekeeper? Order a new marked and mated queen from a reputable bee supplier.

Establishing a Nucleus Hive

A nucleus hive (often called a "nuc"; see Chapter 5) is created by stocking a special miniature hive with a few frames of bees and brood from one of your colonies (see Figure 14-2). Why create a nuc? Some of the reasons include

- A nuc can serve as nursery for raising new queens (see next section).

- A nuc provides you with a handy source of brood, pollen, and nectar to supplement weaker colonies (kind of like having your own dispensary).

- A nuc can be sold to other beekeepers — they're a great way to start a new colony.

- A nuc can be used to populate an observation hive.

- A nuc in the corner of a garden can help with pollination, and be far less maintenance than a regular hive.

The big disadvantage of a nuc is that it doesn't overwinter well. Not enough bees or stored honey is available to see them through the cold months. Combining a nuc with one of your big hives before the winter sets in is best.

Figure 14-2:
A nuc hive doesn't take much room and can be a handy resource for the backyard beekeeper.

Raising your own queens

Dozens of books have been written about raising your own queen bees, so I'll resist dwelling too much on it. Let it suffice to say that raising queens on a small scale can be fun, educational, and even profitable (after all, a single queen sells for $8 to $15). But it's also quite tricky, because it involves a good deal of knowledge about biology and genetics.

A nuc hive is a great housing for raising queens. Besides the housing, you'll also need some special equipment (grafting tools, queen cups, and so on), information on queen rearing, and a good stock of bees from which to raise your queens. My favorite book on the subject is *Queen Rearing and Bee Breeding* by Harry H. Laidlaw Jr. and Robert E Page Jr. (Wicwas Press, Cheshire, Connecticut, 1997, ISBN: 1-878075-08-X).

Starting an Observation Hive

An observation hive is a small hive with a glass panel that enables you to observe a colony of bees without disturbing them or risking being stung. Such hives usually are kept indoors but provide access for the bees to fly freely from the hive to the outdoors (a tube or pipe creates a passage way from the observation hive to outside).

I'm a big believer in having an observation hive — even when you have conventional hives in your garden. The pleasure and added insight they give you about honey bee behavior is immeasurable. Among a few of the rewards you can realize from setting up an observation hive are that it:

- Gives you a barometer on what's happening in a bee colony at any given time of year. That way you can anticipate the needs of your outdoor colonies and better manage your hives. Note that the behavior of bees in an observation hive is influenced by the weather outside (not by the environment indoors).

- Makes possible safe close-up observation of bee behavior. And because you can watch the bees without smoking or opening the hive, the bees' behavior is far more natural. You'll see things that you can never witness while inspecting a conventional hive. I've watched the queen laying eggs, the hive preparing for a swarm, and bees emerging from their cells. I've also studied the bees' remarkable communication dances and much more.

- Provides — because it's kept *indoors* — year-round enjoyment. No need to be a seasonal beekeeper, because you can observe your bees even in the dead of winter.

✔ Serves as a fantastic educational tool for all ages and a stunning conversation piece in your home. Spend endless hours admiring the remarkable world of the honey bee.

✔ Enables you to enjoy the pleasures of beekeeping from the comfort of your home, especially when you don't have the space to keep bees outdoors or can't physically manage a robust outdoor hive.

Observation hives come in all sizes and styles, but practical observation hives for year-round enjoyment are at least three frames thick. Observation hives that are only one or two frames thick don't have enough volume for housing a decent-sized colony that can survive during the winter months. A colony needs ample room to grow and survive on a year-round basis. Furthermore, the bees' behavior is far more natural when they have generous enough space to raise brood and create adequate stores of honey. My favorite design includes three deep frames for brood and three shallow frames for food storage (see Figure 14-3). All observation hives make allowances for feeding the colony sugar syrup (something that you'll have to do on a year-round basis).

A nice six-frame observation hive is available from www.bee-commerce.com, 11 Lilac Lane, Weston, CT 06883. Call 800-748-1911, or 203-222-2268. For detailed information about setting up, maintaining, and using an observation hive, I recommend the book *Observation Hives* by Thomas Webster and Dewey Caron (The A.I. Root Company, Medina, OH, 1999, ISBN: 0-936028-12-2).

Figure 14-3:
The six-frame design of this observation hive means it will overwinter well. The built-in hive-top feeder is esthetically pleasing and a big convenience for the owner.

Planting Flowers for Your Bees

This section was prepared by my friend Ellen Zampino, an avid gardener and an excellent beekeeper.

Flowers and bees are a perfect match. Bees gather nectar and pollen enabling plants to reproduce. In turn, pollen feeds baby bees and nectar is turned into honey to be enjoyed by the bees and you. Everyone's happy.

While many kinds of trees and shrubs are bees' prime source of pollen and nectar, a wide range of flowers contributes to bee development and a bumper crop of honey. You can help in this process by adding some of these flowers to your garden or by *not* removing some that already are there. Did you know that many weeds actually are great bee plants, including the pesky dandelion, clover, goldenrod, and purple vetch? You can grow all kinds of flowering plants in your garden that not only will add beauty and fragrance to your yard but also give bees handy sources of pollen and nectar. You'll hear the warm buzz of bees enjoying them before you even realize the plants are in bloom.

Each source of nectar has its own flavor. A combination of nectars produces great tasting honey. Not all varieties of the flowers described in the sections that follow produce the same quality or quantity of pollen and nectar, but the ones that I list here work well and bees simply love them.

Asters (Aster/Callistephus)

The *Aster family* has more than 100 different species. The aster is one of the most common wildflowers ranging in color from white and pink to light and dark purple. They differ in height from 6 inches to 4 feet and can be fairly bushy. Asters are mostly perennials and blooming times vary from early spring to late fall. However, like all perennials, their blooming period lasts only a few weeks. Several varieties can be purchased as seeds, but you'll also find some aster plants offered for sale at nurseries.

Callistephus are china asters, which run the same range of colors, but produce varied styles of flowers. These pincushions-to-peony style flowers start blooming late in summer and continue their displays until frost. They are annuals. Plants can be bought potted from local nurseries or purchased by seed.

Sunflowers (Helianthus/Tithonia)

Sunflowers are made up of two families. They provide the bees with pollen and nectar. Each family is readily grown from seed, and you may find some nurseries that carry them as potted plants. When you start sunflowers early in the season, make sure that you use peat pots. They are rapid growers that transplant better when you leave their roots undisturbed by planting the entire pot. *Helianthus annuus* include the well-known giant sunflower as well as many varieties of dwarf and multibranched types. Sunflowers no longer are only yellow. They come in a wide assortment of colors, from white to rust and even several varieties of mixed shades. Watch out for the hybrid that is pollenless, because it is of little use to the bees.

Salvia (Salvia/Farinacea-Strata/Splendens)

The Salvia family, with more than 500 varieties, includes the sages *(Salvia officinalis)* and many bedding plants. The sages are good nectar providers. When in bloom, they're covered with bees all day long. The variety of colors and sizes of the *Farinacea* and *Splendens* cover the entire gambit from white, apricot, all shades of red, and purple, to blues with bicolors and tricolors. They're easily found potted in garden stores or available as from seed. *Salvia officinalis* is the sage herb that you can use in cooking.

Bee balm (Monarda)

Bee balm *(Monarda didyma)* is a perennial herb that provides a long-lasting display of pink, red, and crimson flowers in midsummer. They start flowering when they reach about 18 inches and continue to grow to 3 or 4 feet in height. Deadheading them encourages more growth, which can prolong their flowering period. Bee balm is susceptible to powdery mildew but the Panorama type does a good job of fending off this problem. Bee balm is a good source of nectar for bees as well as butterflies and hummingbirds. This family also includes horsemint *(M. punctata),* and lemon mint *(M. citriodora).* The fragrant leaves of most of these plants are used in herbal teas. They are easily found in seed catalogs. Several varieties usually are available at local nurseries.

Hyssop (Agastache)

Anise hyssop *(Agastache foeniculum)* has a licorice fragrance when you bruise its leaves. It produces tall spikes of purple flowers in midsummer. Sometimes you can find a white variety of this plant. The bees happily gather nectar from it. Hyssop flowers from seed the first year that you plant it. Another common hyssop is found in the wild — *Agastache nepetoides.* It has a light, yellowish flower and is found in wooded areas. The seed for this variety are more difficult to find, but some seed houses carry them.

Mint (Mentha)

Chocolate, spearmint, apple mint, peppermint, and orange mint are only a few of the types of mints available. They come in a variety of colors, sizes, fragrances, and appearances, but when they produce a flower, bees are there. Most mints bloom late in the year. Some can be easily grown by seed; other varieties you can start from roots. Mints are easily obtained because they spread readily and many gardeners are happy to share their plants. Most nurseries carry peppermint and spearmint.

Cleome / Spider flower (Cleome)

Spider flower *(Cleome hasslerana)* is heat and drought tolerant and grows well in the cold Northeast. This annual is easy to start from seed and grows more than 4 feet tall with airy flowers that are 6 to 8inches across. It comes in white, pink, and light purple and adds an unusual flower to your garden. It's also a good producer of nectar for the bees, blooming from midsummer to fall.

Thyme (Thymus)

Thyme varieties are low-growing hardy herbs. Common, French, wooly, silver, and lemon are but a few of the varieties available. Several are used in cooking. In spring most nurseries have large selections. These varieties also can be started by seed at least two to three months before planting. Put plants between your steppingstones or at the edges of your garden beds. They bloom from midsummer on. Bees will cover them most of the day gathering nectar, which is aromatic and produces nice tasting honey.

Poppy (Papaver/Eschscholzia)

Danish flag *(Papaver somniferum)*, corn poppy *(P. rhoeas)*, and Iceland poppies *(P. nudicaule)* are easily grown from seed. Some are deep scarlet or crimson, but others are found in pastel shades. All bloom freely from early summer to fall, need full sun, and grow 2 to 4 feet tall. Literature claims that poppies are valuable mostly for the pollen, but I'm sure my bees also are gathering a fair amount of nectar.

California poppies *(Eschscholzia)* are golden orange and easily grown. They are a good pollen source for honey bees. California poppies will self-seed in warmer climates.

Bachelor's buttons (Centaurea)

Annual and perennial selections of bachelor's buttons are available. The annuals *(Centaurea cyanus, C. imperialis),* found in shades of white, pink, yellow, purple, and blue, also are referred to as cornflowers.

The perennial version is a shade of blue that blooms early in summer, and sometimes again in late fall. They're sometimes referred to as mountain blue buttons. Annual and perennial varieties produce an ample supply of nectar. They're easily grown from seed and most nurseries have the annual variety available as potted plants.

Building Your Own Hives

If you're reasonably handy with woodworking, you can build your own hive parts from scratch. Here are some plans to help you along (see Figure 14-4). Remember that precise measurements are critical within a hive. Bees require a precise "bee space." If you wind up with too little space for the bees, they'll glue everything together with propolis. Too much space and they will fill it with burr comb (wax comb built by the bees to fill large voids in the hive). Either way, it makes the manipulation and inspection of frames impossible. So, measure carefully.

Any kind of wood works fine. Commercial makers of bee equipment typically use pine and cypress. These woods are cost effective and easy to work with. But why not give your bees a treat, and spoil them with a hive fashioned from some exotic hardwood? Central American cocobolo or flame mahogany certainly would turn some heads!

Bee space refers to the critical measurement between hive parts that enables bees to freely move about the hive. The space measures ⅜ inch (1 cm).

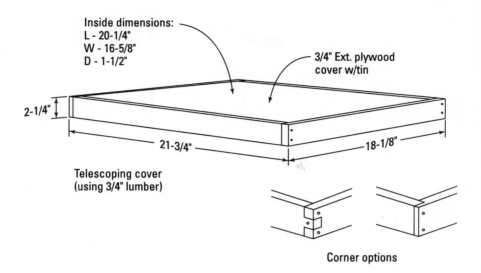

Inside dimensions:
L - 20-1/4"
W - 16-5/8"
D - 1-1/2"

3/4" Ext. plywood cover w/tin

2-1/4"

21-3/4"

18-1/8"

Telescoping cover
(using 3/4" lumber)

Corner options

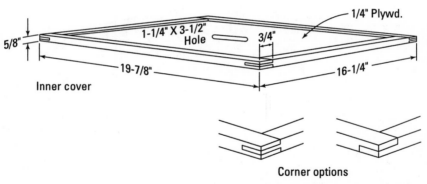

1/4" Plywd.

5/8"

1-1/4" X 3-1/2"
Hole

3/4"

19-7/8"

16-1/4"

Inner cover

Figure 14-4:
These
blueprints
will serve as
a guide if
you decide
to build your
own hive.

Corner options

Courtesy of Barry Birkey and www.beesource.com

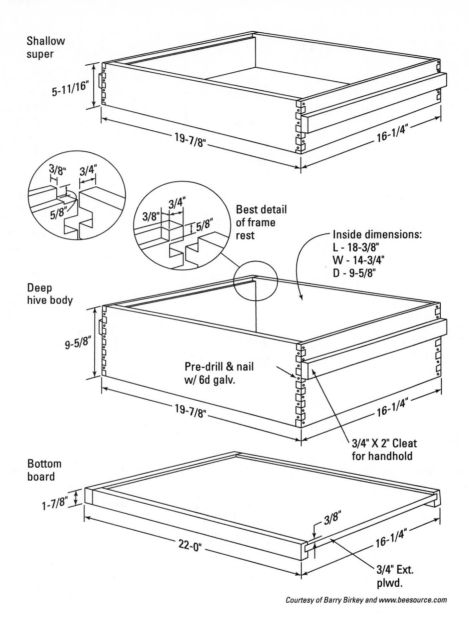

Shallow super

5-11/16"

19-7/8"

16-1/4"

3/8" 3/4"

5/8"

3/4"

3/8" 5/8"

Best detail of frame rest

Inside dimensions:
L - 18-3/8"
W - 14-3/4"
D - 9-5/8"

Deep hive body

9-5/8"

Pre-drill & nail w/ 6d galv.

19-7/8"

16-1/4"

3/4" X 2" Cleat for handhold

Bottom board

1-7/8"

22-0"

3/8"

16-1/4"

3/4" Ext. plwd.

Courtesy of Barry Birkey and www.beesource.com

Check with me at www.bee-commerce.com if you have questions about these blueprints.

Brewing Mead: The Nectar of the Gods

I get restless every winter when I can't tend to my bees. So a number of years ago I looked around for a related hobby that would keep me occupied until spring. I thought, "Why not brew mead?" Mead is a wine made from honey instead of grapes. It was the liquor of the Greek gods and is thought by scholars to be the oldest form of alcoholic beverage. In early England and until about 1600, mead was regarded as the national drink. In fact, the wine that Robin Hood took from Prince John had honey as its base.

When mead is made right, the resulting product is simply delicious! And like a fine, red wine, it gets better and better with age. Many companies supply basic wine- and mead-making equipment to hobbyists (see Figure 14-5). All you need is a little space to set up shop, and some honey to ferment. The key to success is keeping everything sanitary — sterile laboratory conditions!

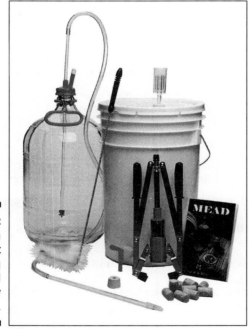

Figure 14-5:
Here's a typical kit for brewing mead (honey wine).

The following recipe produces an extraordinary mead. Technically, this is a *Metheglin,* the term given to mead that is spiced. The recipe yields about 40 bottles of finished product. Adjust the amounts to suit your needs.

Ideally, keep the room's temperature at between 65 and 68 degrees F (the cool basement is a good place to brew mead). If the temperature is higher than 75 degrees F, the yeast may die; if it's less than 50 degrees F, fermentation ceases. Note that a portable space heater with a thermostat helps control basement temperatures during winter.

1. **You must first boil the following honey/water for 30 minutes, skimming off scum as necessary to ensure a finished product that is clear and has better eye appeal. Here is the initial boiling mixture:**

   • 32 pounds of dark wildflower honey

   • 5 gallons of well water (nonchlorinated water)

   • 5 sticks of cinnamon

   • 1 tablespoon cloves

2. **Allow the mixture to cool and then add:**

   • 3½ Tablespoons of acid blend (available at wine-making supply stores)

   • 4¼ Tablespoons of wine yeast nutrient (available at wine-making supply stores)

3. **Pour the mixture into a large (16.5-gallon) initial fermentation tank.**

   Top off with water so that the tank contains a total of 13 gallons of must. *Must* is the water and honey mixture that will ferment into wine. Stir to blend.

4. **Add the following ingredients to the tank of must:**

   • 13 potassium metabisulfite tablets (available at wine-making supply stores) to hinder the growth of undesirable bacteria.

   • 7 drops of antifoam agent (available at wine-making supply stores)

5. **Wait 24 hours, and then add the following to the must in the fermentation tank:**

   • 2½ packets of white wine yeast (stir to blend)

6. **Cover and let the must ferment for one month before performing the first racking.**

   Racking is the process of siphoning off the liquid and leaving the dead yeast cells behind.

7. **After one month, rack liquid into glass carboys. You'll need two or three carboys for this recipe.**

   Fill right up to the neck of the carboy (you want to minimize air space). Place a fermentation valve on each carboy. The valve keeps air and bacteria from entering the carboy. If needed, add potassium metabisulfite tablets to maintain 50 parts per million (ppm).

8. **Rack a total of two or three more times at three-month intervals.**

   Each racking further clears the mead. After the final racking, transfer the mead to sterilized wine bottles and cork tightly. Store bottles on their side. Remember, the longer the mead is aged, the more improved the flavor. Salute!

For more information on making mead, see *Making Mead Honey Wine: History, Recipes, Methods and Equipment* by Roger A. Morse (Wicwas Press, Cheshire, Connecticut, 1992, ISBN: 1878075047).

Create Cool Stuff with Propolis

Propolis (sometimes called "bee glue") is the super-sticky, gooey material gathered by the bees from trees and plants. The bees use this brown goop to fill drafty cracks in the hive, strengthen comb, and to sterilize their home.

Propolis has remarkable antimicrobial qualities that guard against bacteria and fungi. Its use by bees makes the hive one of the most hygienic domiciles found in nature. This remarkable property has not gone unnoticed over the centuries. The Chinese have used it in medicine for thousands of years. Even Hippocrates touted the value of propolis for healing wounds. In addition, propolis has been used for centuries as the basis for fine wood varnishes.

When cold, propolis is hard and brittle. But in warm weather propolis is gummier than words can express. When you inspect your hives at the end of the summer and early autumn (the height of propolis production), you'll discover that the bees have coated just about everything with propolis. The frames, inner cover, and outer cover will be firmly glued together, and they'll require considerable coaxing to pry loose. You'll get propolis all over your hands and clothes, where it will remain for a long, long time. It's a nuisance for most beekeepers. But be sure to take the time to scrape it off or you'll never get things apart next season. Be sure to save the propolis you scrape off with your hive tool! It's precious stuff. I keep an old coffee can in my toolbox and fill it with the propolis I remove from the hive.

Keep a spray bottle of rubbing alcohol in your supply box. Alcohol works pretty well at removing sticky propolis from your hands. But, for goodness sakes, keep propolis off your clothes — because it's nearly impossible to remove.

Many beekeepers encourage the bees to make lots of propolis. Special *propolis traps* are designed just for this purpose. The traps usually consist of a perforated screen that is laid across the top bars — similar to a queen excluder, but the spaces are too narrow for bees to pass through (see Figure 14-6). Instinctively, bees fill all these little holes with propolis. Eventually, the entire trap becomes thickly coated with the sticky, gummy stuff. Remove the trap from the hive (gloves help keep you clean) and place it in the freezer overnight so that the propolis becomes hard and brittle. Like chilled Turkish Taffy, a good whack shatters the cold propolis, crumbling it free from the trap. It then can be used to make a variety of nifty products. I've included some recipes to get you started.

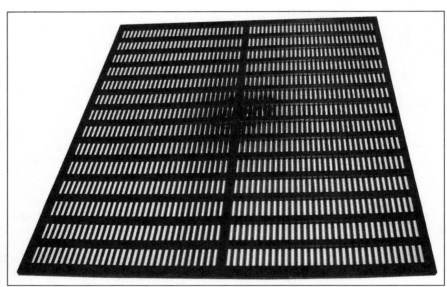

Figure 14-6:
A propolis trap can be placed where the inner cover usually goes. In no time, bees will coat the entire trap with precious propolis.

Propolis tincture

Here's a homemade and all-natural alternative to iodine. **Note:** like iodine, it stains. Use it on minor cuts, rashes and abrasions. Some folks even use a few drops in a glass of drinking water to relieve sore throats. The shelf life of this tincture is several years.

1. **Measure the crumbled propolis and add an equal measure of 100-proof vodka or grain alcohol (for example, one cup propolis and one cup alcohol). Place in an ovenproof bottle with a lid.**

2. **Heat the closed bottle in a 200-degree (Fahrenheit) oven. Shake the bottle every 30 minutes. Continue until the propolis has completely dissolved in the alcohol.**

3. Strain the mixture through a paper coffee filter or a nylon stocking.

4. Bottle the tincture into dropper bottles, which you can get from your pharmacist.

Propolis ointment

This ointment can be applied to minor cuts, bruises and abrasions.

1. **Melt the ingredients in a microwave or in a double boiler.**

   - 1 teaspoon of beeswax

   - 4 teaspoons of liquid paraffin

   - 1 teaspoon of propolis granules

   - 1 teaspoon of honey

2. **Remove from heat and stir continuously until it cools and thickens.**

3. **Pour into suitable jars.**

Propolis varnish

If you happen to have a million-dollar violin made by Stradivarius, you already know that the finest string instruments ever made had a varnish made from propolis. But this superior lacquer need not be reserved for such exclusive uses. Propolis varnish provides a warm, durable finish for any wood project. Here's a recipe from a friend of mine who refinishes museum quality violins.

1. **Combine all ingredients in the following list in a glass jar at room temperature. Cover the jar with a lid. Allow mixture to stand for a week or more while shaking at regular intervals.**

   - 4 parts blond shellac

   - 1 part manila copal (a soft resin)

   - 1 part propolis

2. **Filter solution through a few layers of cheesecloth or a nylon stocking before using.**

Note: The manila copal resin is available from specialty varnish suppliers, such as Joseph Hammerl GmbH & Co. KG, Hauptstrasse 18, 8523 Baiersdorf, Germany.

Making Gifts From Beeswax

Your annual harvest doesn't begin and end with honey. You'll also be collecting plenty of beautiful, sweet-smelling beeswax, which can be cleaned and used for all kinds of artsy projects (see Chapter 13 for instructions on how to clean your wax). You can make candles, furnish polish, and cosmetics for all your friends and neighbors (see Figure 14-7). Better yet, why not sell these goodies at the local farmers' market? Here's some useful information to get you started.

Figure 14-7: Here's a sample of the wonderful products you can make from beeswax.

Beeswax candles

Beeswax candles are desirable; unlike paraffin, they don't drip, don't sputter, and don't smoke, but they do burn a long time. You can make three basic types of candles from beeswax: rolled, dipped, and molded. Buy them in a gift store and they're fantastically expensive. But not when you make them yourself!

Rolled candles

This type of candle is a breeze to make, and no special equipment is required. Thin sheets of craft beeswax are available from candle-making suppliers. The sheets are embossed with a honeycomb pattern (like a sheet of foundation) and come in a huge variety of colors. Purchasing this kind of craft wax probably is best, because it is tough to make without special expensive equipment. To make a rolled candle, place a length of candle wicking along one edge, and roll the sheet up like a jellyroll. Once you get the hang of it, you can easily manufacture several candles in less than a minute. Tie a pair up with a nice ribbon and you have a great gift to take to your next dinner party.

Dipped candles

This is a time consuming process but the end result is beautiful.

1. **Melt beeswax in a tall container (the container can be placed in a hot water bath to keep the wax melted).**

2. **Tie a lead fishing weight to one end of wicking (to make it hang straight) and begin dipping.**

3. **Let each coat of wax cool before dipping again. The more you dip, the thicker the candle becomes.**

With a little finesse, you can create an attractive taper to your dipped candles. You can even add color and scent (your supplier sells what you need, including wicks, coloring dyes, and scents). Elegant!

Molded candles

Candlemaking suppliers offer a huge variety of rubber or plastic molds for candlemaking — from conventional tapers to complex figurines. Just melt your beeswax, pour it into the mold (add color and scent if your want). Don't forget the wick. Let it cool and remove the mold. Easy!

Beeswax furniture polish

My good friend Peter Duncan makes simply beautiful wood furniture. He says that my beeswax wood polish is the finest he's ever used. Smooth enough to apply evenly, beeswax polish feeds and preserves the wood and provides a hard protective finish. Here's my "secret" recipe.

Ingredients:

- ✔ 4 ounces beeswax (by weight)
- ✔ 2 Tablespoons of carnauba wax flakes
- ✔ 2½ cups odorless turpentine or mineral spirits

1. **Melt the waxes in a double boiler.**

2. **Remove the waxes from the heat and stir in the turpentine or mineral spirits.**

3. **Pour into containers (something that looks like a tin of shoe polish is ideal) and let the mixture cool.**

4. **Cover tightly with a lid.**

Apply the polish with a clean cloth and rub in small circles. Turn the cloth as it becomes dirty. Allow the polish to dry, then buff with a clean cloth. If more than one coat is desired, wait two days between applications. This stuff is simply fantastic!

Beeswax cosmetics

Beeswax also makes fantastic, natural cosmetics — which, in turn, make treasured gifts. The process is surprisingly easy, and the ingredients readily available (beeswax from your hive; food-grade oil from the grocery store (extra virgin olive oil works well); and essential oils from the health food store). These recipes from my friend Patty Pulliam can be made firmer by adding more beeswax or softer by adding more oil. However, Patty suggests that the proportions she's listed work nicely for these applications.

Beeswax Lip Balm

For this recipe, you can use any good quality food-grade oil. Patty's favorite is either sweet Almond Oil or Extra Virgin Olive Oil. You can add any edible essential oil to the lip balm, but Patty recommends the tingly freshness of peppermint.

Ingredients:

- 1 part beeswax by dry weight (for example, 1 ounce of beeswax by weight)
- 4 parts food grade oil by liquid volume (for example, 4 ounces of oil by volume)
- A few drops of peppermint oil

1. **In the top of a double boiler, melt the beeswax and oil over medium heat. Stir continuously with a wire whisk until wax has completely melted.**

2. **Remove from the heat and add a few drops of peppermint oil. Continue stirring as the mixture cools and thickens.**

3. **Before it completely cools and sets, pour the warm mixture into small containers.**

4. **Let the balm cool completely before placing the lids on the containers.**

Beeswax Hand Cream

Smooth, creamy and not too oily, this hand cream works better than anything you can buy commercially.

Ingredients:

- 1 part beeswax by dry weight (for example, 1 ounce of beeswax by weight)
- 6 parts olive oil by liquid volume (for example, 6 ounces of oil by liquid volume)
- A few drops of essential oil (such as Lavender, Chamomile, Rose or Lemon Verbena). You may add any essential oil or leave it out entirely to enjoy the natural sweet scent of the beeswax.

1. **In the top of a double boiler, melt the beeswax and oil over medium heat. Stir continuously with a wire whisk until the wax has completely melted.**

2. **Remove from the heat and add a few drops of your favorite essential oil. Continue stirring as the mixture cools and thickens.**

3. **Before it completely cools and "sets," pour the warm mixture into small containers.**

4. **Let the cream cool completely before placing the lids on the jars.**

Chapter 15

Ten Frequently Asked Questions about Bee Behavior

In This Chapter

▶ Getting the answers to questionable situations you may encounter

▶ Understanding *unusual* bee behavior

Having a beekeeping supply business puts me on the receiving end of virtually every question you can possibly imagine about bees and beekeeping. New beekeepers face all kinds of puzzling new situations and concerns every day. I know how gratifying it is for them to have someone they can ask when they just can't seem to figure out what to do next. I had a wonderful mentor when I started beekeeping and it made all the difference when I encountered something baffling.

Not surprisingly, most new beekeepers face the same bewildering situations and ask identical questions. That gives me the illusion of intelligence when I rattle off lucid answers to seemingly impossible conundrums. Actually, it's just that I've had a lot of practice responding to the same questions again and again. The good thing is that I get a little better at it each time.

This chapter includes some of the most frequently asked questions about bee behavior that come my way. Look them over. They may solve a riddle or two for you as you embark on the wonderful adventure of backyard beekeeping.

Help! A million bees are clustered on the front of my hive. They've been there all day and all night. Are they getting ready to swarm?

They're not swarming. Chances are it's hot and humid and the bees are doing just what you'd do — going out on the front porch to cool off. They may spend days and nights outside the hive until the weather becomes more bearable inside. Make sure you've given them a nearby source for water and provided adequate hive ventilation (see Chapter 9).

Is something wrong with my bees? They're standing at the entrance of the hive and it looks like they're just rocking back and forth. Are they sick?

Your bees are fine. They're scrubbing the surface of the hive to clean and polish it. They do this inside and outside the hive. Tidy little creatures, aren't they?

I hived a new package of bees last week. I just looked in the hive. The queen isn't in her cage, and I don't see her or any eggs. Should I order a new queen?

It's probably too early to conclude that you have a problem. Overlooking the queen is easy (she's always trying to run away from the light when you open a hive). Seeing eggs is a far easier method of determining whether you have a queen. But, it may be too soon for you to see eggs. Give it another few days and then look again for eggs. Until they get a better idea of what eggs look like, most new beekeepers have a hard time recognizing them. Have a close look at the images of eggs in the color insert section and in Chapter 2.

A few days after the queen lays the eggs, they hatch into larvae, which are easier to see than eggs. If you see absolutely nothing after ten days (no queen, no eggs, and no larvae), order a new queen from your beekeeping supplier.

Why is my queen laying more than one egg in each cell? Is she just super productive?

Actually you have a problem. More than one egg in a cell means one of two things:

- You have a drone-laying queen.
- You have lost your queen and some of the young worker bees have started laying eggs.

If you have a drone-laying queen, you need to replace her. If you have drone-laying workers, you'll have to remove them from the hive and get a new queen (see Chapter 9). If you don't correct the situation, you'll eventually lose your hive as all the worker bees die off from old age. At that point only drones are left. Without the workers, there will be no bees gathering food, and no workers to feed the helpless drones.

Hundreds of bees are around my neighbor's swimming pool and bird-bath. The bees are creating a problem and the neighbor is blaming me. What can I do?

Bees need lots of water in summer, and your neighbor's pool and birdbath are probably the bees' closest sources. You must provide your bees with a closer source of water (see the sidebar in Chapter 3). If they're already imprinted on your neighbor's oasis, you may have to "bait" your new water source with a light mixture of sugar water. After the bees find your sweet new watering hole, you can switch to 100 percent water.

A tremendous amount of activity is present at the entrance of the hive. It looks like an explosion of bees flying in and out of the hive. The bees seem to be wrestling with each other and tumbling onto the ground. They appear to be fighting with each other. What's going on?

It sounds like you have a robbing situation. Your bees are trying to defend the hive against invading bees that are stealing honey from your hive. You must call a halt to this activity before the robbing bees steal all the honey and many bees die in the battle (see Chapter 9 for tips on how to prevent and solve this situation).

My bees had been so sweet and gentle, but now I'm scared to visit the hive. They have become unbearably aggressive. What can I do?

Bees become more aggressive for a number of different reasons. Consider the following possibilities, and see whether any apply to your situation:

- ✔ A newly established colony always starts out gentle. As the colony grows in size and the season progresses, the bees become more protective of their honey stores. Likewise, a growing colony means many more bees for you to deal with. But if the colony is handled with care this is seldom a problem. Be gentle as you work with your colony.

- ✔ Incorrect use (or lack of use) of the smoker can result in irritable colonies. See Chapter 6 for information about how to use your smoker.

- ✔ Do you launder your bee clothes and veil? Previous stings on gloves and clothing can leave behind an alarm pheromone that can stimulate defensive behavior when you revisit the hive. Be sure to keep your garments clean. You can also smoke the area of the sting to disguise any alarm pheromone that may linger on clothing or on your skin.

- ✔ When colonies are raided at night by skunks or other pirates, they can become cross and difficult to deal with. See Chapter 11 for ideas on how to remedy these situations.

- ✔ Do you still have your original queen? Are you sure? If you had a marked queen you'd know for certain whether the queen now heading your colony is your original queen (see if she's marked!). A colony that supersedes the queen sometimes can result in more aggressive bees. That's because you have no guarantee of the new genetics. The new queen mated with drones from goodness knows where. Her offspring may not be as nice as the carefully engineered genetics provided by your bee supplier. When this happens, order a marked and mated queen from your supplier to replace the queen that is now in your hive.

When you purchase a marked queen from a supplier, the marking stays on for the full life of the queen. It's like spilling nail polish on the living room carpet. It never wears off!

I see white spots on the undersides of my bees. I'm worried these might be mites or some kind of disease. What are these white flecks?

This isn't a problem. The white flakes that you see are bits of wax produced by glands on the underside of the bee's abdomen. They use this wax to build comb. All is well.

The bees have carried dead larvae out of the hive and dumped them in and around the entrance of the hive. What's going on?

Bees remove any dead bees and larvae from the hive. They keep a clean house. The dead larvae may be *chilled brood,* or brood that died when the temperature took a sudden and unexpected drop. Larvae that look hard and chalky may be a sign of chalkbrood (see Chapter 10 for more information on chalkbrood). Either case is fairly commonplace. You don't need to be concerned unless the number of dead bees and larvae is high (more than ten).

It's mid-winter and I see quite a few dead bees on the ground at the hive's entrance. Is this normal?

Yes. Seeing a few dozen dead bees in and around the hive's entrance during the winter months is normal. The colony cleans house on mild days and attempts to remove any bees that have died during the winter. In addition, some bees may take "cleansing flights" on mild sunny days but may become disoriented or caught in a cold snap. When that's the case, they don't make it all the way back to the hive — dropping dead in the snow. Seeing *more* than a few dozen dead bees may be an indication of a health problem, so it may be time for a closer inspection on the first mild, sunny day.

Chapter 16

My Ten Favorite Honey Recipes

In This Chapter

▶ Using honey in place of sugar in recipes

▶ Baking and cooking with honey

*T*here's a good chance that you'll be able to harvest 100 pounds or more of honey from each of your hives. That's a lot of honey. Unless you eat a whole lot of toast, you may want to consider other ways to utilize your copious crop (see Figure 16-1). Honey is not only wholesome, delicious, sweet, and fat-free, but it's also incredibly versatile. You'll find uses for honey in a myriad of recipes that call for a touch of sweetness.

In this chapter, I include ten of my favorite recipes from the National Honey Board. For many additional recipes, be sure to visit their Web site (www.honey.com) or write them at National Honey Board, 390 Lashley Street, Longmont, CO 80501-6045.

Before I jump into the recipes themselves, here are some tips for cooking with honey:

✔ Because of its high fructose content, honey has a higher sweetening power than sugar. This means you can use less honey than sugar to achieve the desired sweetness.

✔ To substitute honey for sugar in recipes, start by substituting up to half of the sugar called for. With a little experimentation, honey can replace all the sugar in some recipes.

✔ When measuring honey, keep in mind that one 12-ounce jar of honey equals a standard measuring cup.

✔ For easy cleanup when measuring honey, coat the measuring cup with nonstick cooking spray or vegetable oil before adding the honey. The honey will slide right out.

✔ In baking, honey helps baked goods stay fresh and moist longer. It also gives any baked creation a warm, golden color. When substituting honey for sugar in baked goods, follow these guidelines:

  • Reduce the amount of liquid in the recipe by ¼ cup for each cup of honey used.

  • Add ½ teaspoon of baking soda for each cup of honey used.

  • Reduce the oven temperature by 25 degrees to prevent overbrowning.

Figure 16-1:
Honey goes with all sorts of foods!

Courtesy of National Honey Board

Honey Lemonade with Frozen Fruit Cubes

Yield: *6 servings*

1½ cups lemon juice

¾ cup honey

9 cups water

48 pieces of assorted fruit

1 Combine the lemon juice and honey in a large pitcher; stir until the honey is dissolved. Stir in the water.

2 Place 1 or 2 pieces of fruit in each compartment of 2 ice cube trays. Fill each compartment with honey lemonade and freeze. Chill the remaining lemonade.

3 To serve, divide the frozen fruit cubes among tall glasses and fill with the remaining lemonade.

Per serving (1½ cups): Calories 164 (1% of Calories from Fat); Total Fat <1g; Protein <1g; Carbohydrates 45g; Cholesterol 0mg; Sodium 3mg; Dietary Fiber 1g

Banana Yogurt Shake

Yield: *4 servings*

1½ cups 2% low-fat milk

2 ripe bananas, peeled

1 cup low-fat plain yogurt

¼ cup honey

1 teaspoon vanilla extract

½ teaspoon ground cinnamon

Dash ground nutmeg

5 ice cubes

Combine all the ingredients except the ice cubes in a blender or food processor. Process until thick and creamy. With the motor running, add the ice cubes and process until smooth. Pour into tall glasses and serve.

Per serving (1 cup): Calories 232 (22% of Calories from Fat); Total Fat 6g; Protein 6g; Carbohydrates 40g; Cholesterol 18mg; Sodium 82mg; Dietary Fiber 1g

Spiced Tea

Yield: 4 servings

4 cups freshly brewed tea

¼ cup honey

4 cinnamon sticks

4 whole cloves

4 orange slices

Combine the tea, honey, cinnamon sticks, and cloves in a large saucepan and simmer for 5 minutes. Serve hot, or let cool and serve with ice cubes for a great iced tea treat! Garnish each cup with an orange slice.

Per serving (1 cup): Calories 76 (1% of Calories from Fat); Total Fat <1g; Protein <1g; Carbohydrates 20g; Cholesterol 0mg; Sodium 8mg; Dietary Fiber <1g

Apricot Honey Bread

Yield: 12 servings

3 cups whole-wheat flour

3 teaspoons baking powder

1 teaspoon ground cinnamon

½ teaspoon salt

¼ teaspoon ground nutmeg

1¼ cups 2% low-fat milk

1 cup honey

1 egg, lightly beaten

2 tablespoons vegetable oil

1 cup chopped dried apricots

½ cup chopped almonds or walnuts

½ cup raisins

1 Combine the flour, baking powder, cinnamon, salt, and nutmeg in a large bowl and set aside. Combine the milk, honey, egg, and oil in separate large bowl.

2 Pour the milk mixture over the dry ingredients and stir until just moistened. Gently fold in the apricots, nuts, and raisins.

3 Pour into a greased 9-x-5-x-3-inch loaf pan. Bake at 350 degrees for 55 to 60 minutes or until a wooden pick inserted near the center comes out clean.

Per serving: Calories 302 (15% of Calories from Fat); Total Fat 6g; Protein 7g; Carbohydrates 61g; Cholesterol 20mg; Sodium 154mg; Dietary Fiber 5g

Honey Pumpkin Pie

Yield: *8 servings*

1 can (16 ounces) solid pack pumpkin

1 cup evaporated low-fat milk

¾ cup honey

3 eggs, lightly beaten

2 tablespoons all-purpose flour

1 teaspoon ground cinnamon

½ teaspoon ground ginger

½ teaspoon dark rum or rum extract

Pastry for single 9-inch piecrust

Combine all the ingredients except the pastry in a large bowl; beat until well blended. Pour into the pastry-lined 9-inch pie plate. Bake at 400 degrees for 45 minutes or until a knife inserted near the center comes out clean.

Per serving (1 slice): *Calories 284 (27% of Calories from Fat); Total Fat 9g; Protein 7g; Carbohydrates 46g; Cholesterol 82mg; Sodium 209mg; Dietary Fiber 2g*

Honey Barbecue Baste

Yield: *1 cup*

1 tablespoon vegetable oil

¼ cup minced onion

1 clove garlic, minced

1 can (8 ounces) tomato sauce

⅓ cup honey

3 tablespoons vinegar

2 tablespoons dry sherry

1 teaspoon dry mustard

½ teaspoon salt

¼ teaspoon freshly ground black pepper

1 Heat the oil in a saucepan over medium heat until hot. Add the onion and garlic; cook and stir until the onion is tender.

2 Add the remaining ingredients and bring to a boil; reduce the heat to low and simmer for 20 minutes. Serve over grilled chicken, pork, spareribs, salmon, or hamburgers.

Per serving (about 3 tablespoons): *Calories 103 (21% of Calories from Fat); Total Fat 3g; Protein 1g; Carbohydrates 19g; Cholesterol 0mg; Sodium 408mg; Dietary Fiber 1g*

Honey-Mustard Roasted Potatoes

Yield: *4 servings*

4 large baking potatoes (about 2 pounds)

½ cup Dijon mustard

¼ cup honey

½ teaspoon crushed dried thyme leaves

Salt and pepper to taste

1 Peel the potatoes and cut each one into 6 to 8 pieces. Cover the potatoes with salted water in a large saucepan and bring to a boil over medium-high heat. Cook the potatoes for 12 to 15 minutes or until just tender. Drain.

2 Combine the mustard, honey, and thyme in a bowl. Toss the potatoes with the mustard mixture until evenly coated. Arrange the potatoes on a foil-lined baking sheet coated with nonstick cooking spray. Bake at 375 degrees for 20 minutes or until the potatoes begin to brown around the edges. Season to taste with salt and pepper.

Per serving: Calories 296 (6% of Calories from Fat); Total Fat 2g; Protein 6g; Carbohydrates 65g; Cholesterol 0mg; Sodium 726mg; Dietary Fiber 3g

Honey Herb Salad Dressing

Yield: *½ cup*

¼ cup white wine vinegar

¼ cup honey

2 tablespoons chopped fresh basil or mint

1 tablespoon minced green onion

Salt and pepper to taste

Combine all ingredients in a small bowl and whisk briskly until mixed well.

Per serving (2 tablespoons): Calories 81 (0% of Calories from Fat); Total Fat 0g; Protein 0g; Carbohydrates 21g; Cholesterol 0mg; Sodium 4mg; Dietary Fiber <1g

Grilled Honey Garlic Pork Chops

Yield: *4 servings*

¼ cup lemon juice

¼ cup honey

2 tablespoons soy sauce

1 tablespoon dry sherry

2 cloves garlic, minced

4 boneless center-cut pork chops, about 4 ounces each

1 Combine the lemon juice, honey, soy sauce, sherry, and garlic in a bowl. Place the pork chops in a shallow baking dish and pour the marinade over the pork. Cover and refrigerate for 4 hours or overnight.

2 Remove the pork from the marinade. Heat the remaining marinade to a simmer in a small saucepan over medium heat. Grill the pork over medium-hot coals for 12 to 15 minutes, turning once during cooking and basting frequently with marinade. A meat thermometer should register 155 to 160 degrees when inserted into the pork.

Per serving: Calories 248 (26% of Calories from Fat); Total Fat 7g; Protein 25g; Carbohydrates 20g; Cholesterol 61mg; Sodium 604mg; Dietary Fiber <1g

Linguini with Honey Prawns

Yield: *4 servings*

1 pound prawns, peeled and deveined

½ cup julienned carrots

½ cup sliced celery

½ cup diagonally sliced green onions

3 cloves garlic, minced

2 tablespoons olive oil

½ cup water

¼ cup honey

4 teaspoons cornstarch

1 teaspoon salt

¼ teaspoon crushed red pepper flakes

¼ teaspoon crushed dried rosemary leaves

1 pound cooked linguini (al dente), kept warm

1 In a large, heavy skillet over medium-high heat, stir-fry the prawns, carrots, celery, green onions, and garlic in the oil for about 3 minutes, or until the prawns start to turn pink.

2 Combine the water, honey, cornstarch, salt, red pepper flakes, and rosemary in a small bowl and mix well. Add to the prawn mixture and stir-fry for about 1 minute or until the sauce thickens a bit. Pour over the cooked pasta and serve immediately.

Per serving: Calories 445 (19% of Calories from Fat); Total Fat 10g; Protein 30g; Carbohydrates 61g; Cholesterol 172mg; Sodium 737mg; Dietary Fiber 4g

Appendix A:
Helpful Resources

A s a new beekeeper, you'll welcome all the information that you can get your hands on. In this chapter I present a bunch of resources that I find mighty useful: Web sites, vendors, associations, and journals.

Honey Bee Web Sites

I confess. I've become an Internet addict! What in the world did I do before cyberspace? For one thing, I took a lot of trips to the library. But not even the most determined library search of years gone by would have turned up the plethora of bee-related resources that are only a click away on the Web. Just enter the word "beekeeping" or "honey bees" into any of the search engines, and you'll come up with hundreds (even thousands) of finds. Like all things on the Net, many of these sites tend to come and go. A few are outstandingly helpful. Some are duds. Others have ridiculous information that may lead you to trouble. So be careful as you browse around in cyberspace. But, by all means, poke around the Web and see what you come up with. Here are some of my tried-and-true bee-related favorites. Each is worth a visit.

Bee-commerce.com

www.bee-commerce.com

In the spirit of shameless self-promotion, I include my own Web site. Actually, with more than 6 million hits a year, it's, indeed, a popular beekeeping site, featuring many free "how-to" downloads, instruction sheets, and a wide array of beekeeping supplies, equipment, startup kits, books, and videos. My colleagues and I always are available to answer questions and provide beekeeping advice. So be sure to visit bee-commerce and feel free to send me an e-mail. I'd love to hear from you!

See the coupon at the end of this book for a special discount offer from bee-commerce.com.

About.com (beekeeping homepage)

`about.com`

Go to about.com, a slick Web site, and search on "beekeeping." You can find just about every bee resource that you can possibly imagine. If you're looking for *anything* to do with bees and beekeeping, this is a great place to start. A professional guide who is carefully screened and trained by the hosting company runs the site. This is a good Web site to bookmark.

Apiservices — Virtual beekeeping gallery

`www.apiservices.com`

This European site is a useful gateway to scores of other beekeeping sites: forums, organizations, journals, vendors, conferences, images, articles, catalogues, apitherapy, beekeeping software, plus much more. It can be accessed in English, French, Spanish, and German and is nicely organized.

BeeHoo — The beekeeping directory

`www.beehoo.com`

This comprehensive international site has many helpful articles, information sheets, instructional guides, photos, and links of interest for the backyard beekeeper. The site is viewable in English or in French and is definitely worthy of a bookmark.

Bee-Source.com

`www.beesource.com`

This site includes a nicely organized collection of bee-related articles, resources, and links, and it features sections on bees in the news, editorials, an online book store, a listing of beekeeping suppliers, plans for building your own equipment, discussion groups, bulletin boards, and much more.

CyberBee

```
www.cyberbee.net
```

This site is maintained by a beekeeper and contains many helpful references to bee-related information. It also has helpful links for finding local bee clubs, state bee inspectors, and other regional bee-related resources.

Mid Atlantic Apiculture Research and Extension Consortium (MAARAC)

```
maarec.cas.psu.edu/index.html
```

This research and extension consortium is packed with meaningful information for beekeepers worldwide. Download extension publications, find out more about videos, slide shows, software, courses that are available from the organization, and read about honey bee research currently underway. You can also discover important local beekeeping events planned in the Mid-Atlantic Region and other national and international meetings of importance to beekeepers. The USDA/ARS (Beltsville Bee Lab) is an active member of MAARAC. This site is more techie than others.

Mite control using essential oils

```
www.wvu.edu/~agexten/varroa.htm
```

This site includes useful information about using essential oils and menthol for the natural control of parasitic mites. If you want to avoid using chemical pesticides in your hives, this information site is a must-read. This text-only site is administered by Dr. Jim Amrine, an entomologist and acarologist at West Virginia University.

National Honey Board

```
www.nhb.org
```

This nonprofit government agency supports the commercial beekeeping industry. The folks at NHB are enormously helpful and accommodating. The well-designed site is a great source for all kinds of information about honey. You'll find articles, facts, honey recipes, and plenty of beautiful images — all available as free downloads. Well worth the visit!

Bee Organizations and Conferences

Here are ten of my favorite national and international beekeeping associations. Joining one or two of these is a great idea, because their newsletters alone are worth the price of membership (dues are usually modest). Most of these organizations sponsor meetings and conferences. On the agenda: bees, bees, and more bees. Attending one of these meetings (even if you only do it occasionally) is a fantastic way to learn about new tricks, find new equipment, and meet some mighty nice people with similar interests.

American Apitherapy Society

www.apitherapy.org

This nonprofit organization researches and promotes the benefits of using honey bee products for medical use. A journal published by the society four times a year. Once a year, AAS organizes a certification course.

American Apitherapy Society
5390 Grande Road
Hillsboro, OH 45133
Telephone: 937-364-1108
Fax: 937-364-9109
Email: aasoffice@in-touch.net

American Bee Breeder's Association

The American Bee Breeder's Association is an organization with a mission to support the interests and needs of professional bee breeders.

Fred Rossman
GA Hwy 33 North, PO 909
Moultrie, GA 31776-0909
Toll free: 1-800-333-7677
Telephone: 912-985-7200
Fax: 912-739-4821
Email: jrossman@surfsouth.com.

American Beekeeping Federation

www.abfnet.org

This nonprofit organization plays host to a large beekeeping conference and trade show each year. The meetings are worth attending because they include a plethora of interesting presentations on honey bees and beekeeping. By all means join this organization to take advantage of its bimonthly newsletter. The organization's primary missions are benefiting commercial beekeepers and promoting the benefits of beekeeping to the general public. Commercial beekeepers dominate the membership; however, the hobbyist will find value in membership as well.

American Beekeeping Federation
PO 1038
Gesup, GA 31598
Telephone: 912-427-4233
Email: info@abfnet.org.

American Honey Producers Association

www.americanhoneyproducers.org

The American Honey Producers Association is a nonprofit organization dedicated to promoting the common interest and general welfare of the American Honey Producer. The handsome Web site provides the public and other fellow beekeepers with industry news, membership information, convention schedules, cooking tips, and contact information.

American Honey Producers Association
Route 3, PO 258
Alvin, TX 77511
Telephone: 713-992-0802
Fax: 713-996-9484

Apimondia: International Federation of Beekeepers' Associations

www.apimondia.org

Apimondia is a huge international organization comprised of 55 national beekeeping associations from 49 countries, representing more than 5 million members. The organization plays host to a large international conference and trade shows every other year.

Apimondia
Corso Vittorio Emanuele 101
I-00186 Rome, Italy
Telephone: ++39 066852286
Email:

Bee Research Laboratory

www.barc.usda.gov/psi/brl/brl-page.html

Known as "Beltsville" in the bee world, the Bee Research Laboratory is a division of the U.S. Department of Agriculture and a good agency to know about. After all, if you're an American, your tax dollars are paying for it! It's the oldest of the federal bee labs. Launched in 1891, this is one of the leading research labs in the country. The list of scientists who have worked at Beltsville in the past reads like a "Who's Who of American Beekeeping Research." If you ever needed to (let's hope not), you can send samples of your sick bees to the lab for analysis. The lab also is consulted when there's a question about whether a colony is Africanized. The lab's Web site has a lot of helpful information on bee diseases, and a bunch of other technical stuff.

Bee Research Laboratory
Bldg. 476 BARC-East
Beltsville, MD 20705
Telephone: 301-504-8205
Fax: 301-504-8736
E-mail: FeldlauM@ba.ars.usda.gov

Eastern Apiculture Society

www.easternapiculture.org

The Eastern Apiculture Society (EAS) was established in 1955 to promote honey bee culture, the education of beekeepers, and excellence in bee research. Membership consists mostly of beekeepers east of the Mississippi River. Every summer, EAS conducts its annual conference in one of its 22 member states/provinces. About 500 people, from around the world attend this conference every year. The event is simply wonderful for a beekeeper. You can even take a comprehensive exam to become certified as an EAS "master beekeeper." By all means, try to attend one of these weeklong adventures. You'll discover new products and techniques, learn a ton of stuff, and make some lifelong beekeeping friends. EAS also publishes a quarterly newsletter.

Eastern Apiculture Society
PO 300
Essex, NY 12936
Telephone: 518-963-7593
Email: eas@willex.com.

International Bee Research Association

www.cf.ac.uk/ibra/

Founded in 1949, the International Bee Research Association (IBRA) is a non-profit organization with members in almost every country in the world. Its mission is to increase awareness of the vital role of bees in agriculture and the natural environment. The organization is based in the United Kingdom. IBRA publishes several journals and sponsors international beekeeping conferences. Lots of good information and bee-related links can be found on the IBRA Web site.

International Bee Research Association
18 North Road
Cardiff, CF10 3DT, UK
Telephone: +44 (0) 29 20 372409
Fax: +44 (0) 29 20 665522
Email:

The Western Apiculture Society

A small association of beekeepers mostly from the western part of U.S. and Canada, the Western Apiculture Society offers a quarterly newsletter and an annual conference.

Eric Mussen
Dept. of Entomology, University of California Davis
Davis, CA 95616
Telephone 530-752-0472
Email: ecmussen@ucdavis.edu.

Apiary Inspectors of America

www.mda.state.mn.us/AMS/apiary/aiahome.htm

The Apiary Inspectors of America is an organization that promotes better beekeeping conditions in North America through more effective laws and methods for the suppression of bee diseases and by encouraging a mutual understanding and cooperation between apiary inspection officials.

Blane White, Supervisor
Apiary Inspection
Minnesota Department of Agriculture
90 West Plat Boulevard
Saint Paul, MN 55107
Telephone: 651-296-0591
Email: blane.white@state.mn.us

Bee Journals & Magazines

Are you ready to curl up with a good article about honey bees? A bunch of publications are worth a read. Subscribing to one or more of them provides you with ongoing sources of useful beekeeping tips and practical information. And the ads in these journals are a great way to learn about new beekeeping toys and gadgets. Here are ten English language journals of interest.

American Bee Journal

www.dadant.com/journal

The *American Bee Journal* has been around for more than a century. Although it primarily targets the professional beekeeper, backyard beekeepers value its articles on practical beekeeping techniques and beekeeping news from around the world.

Dadant & Sons, Inc.
51 South Second Street
Hamilton, IL 62341
Toll free: 1-800-637-7468
Telephone: 217-847-3324
Fax: 217-847-3660

Australian Bee Journal

The *Australian Bee Journal* is a monthly magazine for beekeepers with a focus on beekeeping in Victoria. Keeping abreast of beekeeping in other countries always is interesting. Often you can discover new techniques and medication protocols that haven't yet reached our shores.

RSD McKenzies Hill,
Castlemaine, Victoria 3450
Australia
Telephone: (03) 5472 2161
Fax: (03) 5472 3472

Bee Culture

www.beeculture.com

This easy-to-digest journal has been around since the late 1800s. Articles are aimed at the needs and interests of the backyard beekeeper and small-scale honey producers. It features a wide range of "how-to" articles, Q&A, honey recipes, and industry news. Hands down, this is the "bible" for the hobbyist beekeeper.

Every year, *Bee Culture* magazine publishes "Who's Who in North American Beekeeping." This article lists many regional bee clubs, agencies, and organizations. It's a great resource for finding other beekeepers. The Web address above contains an online version of this database that enables you to search for bee clubs in your area.

Joining a local bee club/association is an excellent way to latch onto a mentor. Clubs usually schedule meetings during the year with guest speakers on different topics. A ton of local beekeeping clubs and associations can be found across the U.S. — far too many to list here. Besides *Bee Culture*'s list of bee clubs, you can contact your local county agricultural extension office to find out about bee clubs in your area.

A.I. Root Company
623 W. Liberty St.
Medina, OH 44256
Toll free: 1-800-289-7668, Ext. 3220
Fax: 330-725-5624

See the coupon at the end of this book for a special subscription discount to *Bee Culture*.

Bee World

Bee World is a quarterly journal that digests research studies and articles from around the world. Reliable and practical information come from bee experts worldwide. It's published by IBRA.

www.cf.ac.uk/ibra/journal2.shtml

International Bee Research Association
18 North Road
Cardiff CF1 3DY United Kingdom
Telephone: +44 (0) 29 20 372409
Fax: +44 (0) 29 20 665522

British Bee Journal

The *British Bee Journal* is one of the leading bee journals in the United Kingdom. It provides useful insight into the English way of beekeeping, where they've been keeping bees for centuries.

British Bee Journal
46 Queen Street
Geddington NR Kettering,
Northhamptonshire NN14 1 AZ, United Kingdom

Canadian Beekeeping

For more than 20 years, *Canadian Beekeeping* has served beekeepers across Canada, bringing them news, interviews, and featured articles. This journal provides plenty of information on seasonal management, new methods, equipment, treatments, and it reports on conferences, competitions, and regulations.

www.canadianbeekeeping.com

Canadian Beekeeping
POB 678
Tottenham, Ontario
Canada L0G 1W0
Telephone: 905-936-4975
Email: editor@canadianbeekeeping.com

Scottish Bee Journal

From the land of haggis comes the *Scottish Bee Journal,* which is devoted to beekeeping in Scotland. I find it interesting and helpful to keep abreast of methods and techniques from other countries. You can always find out something new from other beekeepers.

Scottish Bee Journal
R.N.H. Skilling
FRSA, 34 Rennie Street
Kilmarnock, Ayrshire
Scotland

South African Bee Journal

Beekeeping in South Africa has its challenges — given that Africa is home to the infamous "killer bee." The *South African Bee Journal* has many interesting articles about keeping bees in warm climates.

South African Bee Journal
POB Box 41
Moderfontein 1645
South Africa

The New Zealand Beekeeper

The *New Zealand Beekeeper* is published 11 times a year by the National Beekeeping Association of New Zealand. This journal offers excellent articles, rich with useful information.

The New Zealand Beekeeper
Farming House
211-213 Market Street South
POB 307, Hastings
New Zealand

The Speedy Bee

The primary focus of *The Speedy Bee* is on business topics for commercial beekeepers, including news of U.S. trade regulations, regulations dealing with pesticide use, and quarantines that impact honey bees. Publication of this newspaper can be sporadic. To subscribe, write:

The Speedy Bee
POB 998
Jesup, GA 31545

Beekeeping Supplies & Equipment

Where do you find all the neat stuff you need to become a beekeeper? Where do you buy bees? You can start by taking a look in the local yellow pages under "beekeeping supplies." Maybe you'll get lucky and find a listing for a local beekeeper who sells supplies out of his or her garage. That's kind of cool, because it gives you face-to-face access to your own personal mentor. Alternatively, you can deal directly with one of the major bee suppliers. They all offer mail-order catalogs and many have secure e-commerce enabled Web sites. Some provide online advice (your online mentor). I've listed ten of the more popular suppliers.

Bee-commerce.com

www.bee-commerce.com

Okay. This is *my* company, so yes I have a big bias here. Of course I think it's the best online bee supply business! Why? Because bee-commerce.com truly offers the highest quality beekeeping supplies and equipment. No junky stuff is found on this site. The company offers secure e-commerce shopping and personalized support designed exclusively for the backyard beekeeper. A free download section is provided with helpful instruction sheets and articles. Most products are shipped the same day the order is placed. And with a 100 percent satisfaction guarantee, customers can order with complete confidence. Order online or by phone, and speak to a real person 24 hours a day, 365 days of the year. The company's beekeeping experts happily serve as your online mentors. The site offers mail order and a user-friendly e-commerce interfaces. Bee-commerce also sells package bees and queens.

Bee-commerce, WoodsEnd, Inc.
11 Lilac Lane
Weston, CT 06883
Toll Free: 1-800-784-1911
Telephone: 203- 222-2268
Fax: 413-653-1978
Email: info@bee-commerce.com.

See the coupon at the end of this book for a special discount offer from bee-commerce.com.

The Beez Neez Apiary Supply

www.beezneezapiary.com

Founded in 1993, the Beez Neez Apiary Supply provides beekeeping supplies and equipment by mail order and e-commerce. It also offers candlemaking supplies, honey candy, and mead making kits. An online catalog is available.

The Beez Neez Apiary Supply
403-A Maple Avenue
Snohomish, WA 98290-2562
Telephone: 360-568-2191
Email: jean@beezneezapiary.com

Brushy Mountain Bee Farm

www.beeequipment.com/

Brushy Mountain Bee Farm features a large selection of beekeeping supplies and equipment by mail order and e-commerce. Brushy Mountain manufacture nice quality hives. Its Web site is attractive but a little awkward when it comes to finding stuff.

Brushy Mountain Bee Farm
610 Bethany Church Road
Moravian Falls, NC 28654
Toll Free: 1-800-BEESWAX
Fax: 336-921-2681
Email: sales@beeequipment.com.

Dadant & Sons, Inc.

www.dadant.com

Dadant & Sons, Inc., provides beekeeping supplies and equipment by mail order and e-commerce through its ten regional offices around the U.S. One of the largest suppliers in the U.S., it has been in business for more than 135 years. Dadant & Sons' primary business is selling decorative and religious beeswax candles.

Dadant & Sons Inc.
51 South Second
Hamilton, IL 62341
Telephone: 217-847-3324
Fax: 217-847-3660
Toll Free: 1-800-637-7468
Email: Dadant@dant.com

Glorybee Foods, Inc.

www.glorybee.com/

Beekeeping equipment plus soap, skin care, aromatherapy, and candle-making supplies are available by mail order and e-commerce from Glorybee, Inc. Founded in 1978, it also is a good resource for unique bee-related gifts.

Glorybee Foods, Inc.
120 North Seneca Road
PO Box 2744
Eugene, OR 97402
Toll Free: 1-800-456-7923
Fax: 541-689-9692
Email: sales@glorybee.com.

Mann Lake Ltd

www.mannlakeltd.com

Mann Lake, Ltd., offers beekeeping supplies, equipment, and medication by mail order and e-commerce. Its Web site features an online catalog and a bee-keeping learning center.

Mann Lake, Ltd
501 First Street
Hackensack, MN 56452
Toll free: 1-800-880-7694
Fax: 218-675-6156
Email: beekeepr@mannlakeltd.com.

Rossman Apiaries

www.gabees.com

Rossman Apiaries is a family business that features beekeeping supplies and equipment by mail order only. It specializes in nice cypress woodenware. Rossman also manufactures special-order items for beekeepers.

Rossman Apiaries
GA Highway 33 North
PO Box 909
Moultrie, Georgia 31776-0909
Toll Free: 1-800-333-7677.

Swienty Beekeeping Equipment

www.swienty.com

European supplier Swienty Beekeeping Equipment offers beekeeping supplies by mail order and fax only, including a nice selection of unique products not readily available in the U.S. Its Web site is published in four languages.

Hørtoftvej 16 - DK-6400
Sønderborg, Denmark
Telephone: ++ 74486969
Fax: 74488001
Email: shop@swienty.com

Thorne Beekeeping Supply

www.thorne.co.uk

Thorne Beekeeping Supply offers beekeeping supplies and equipment by mail order only. The company has three store locations, all based in the United Kingdom. It has plenty of stuff on its Web site, but the site is a little cluttered and difficult to negotiate.

E.H. Thorne, Ltd.
Beehive Works
Wragby, Market Rasen
LN3 5LA, UK
Telephone: 44+(0)1 673 858 555
Fax: 44+(0)1 673 857 004

The Walter T. Kelley Company

www.netgrab.com/kelleycompany/

The Walter T. Kelley Company provides beekeeping supplies and equipment by mail order only. The grade of their products tends to appeal to budget-conscious shoppers.

Walter T. Kelley Company
P.O. Box 240
3107 Elizabethtown Road
Clarkson, KY 42726-0240
Telephone: 270-242-2012
Toll Free: 1-800-233-2899
Fax: 270-242-4801
Email: kelleybees@creative-net.net

Appendix B:
Beekeeper's Checklist

• •

Hive number/location _____

Date of this inspection _____

Date queen/hive was established _____

Use this simple checklist every time that you inspect your bees. It's okay to photocopy it. Be sure to date copies and keep them in a loose-leaf notebook for future reference and comparison. Use one form for each of your hives. For more details on what to look for during inspections, be sure to read Chapters 7 through 11. Happy beekeeping!

Observations	*Notations*
❑ Observe bees at entrance. (Look for dead bees or abnormal behavior and appearance.)	
❑ Do you see "spotting" of feces on the hive? (If yes, the bees may have Nosema and need to be medicated.)	
❑ What is the condition of your equipment? (Note any needed repairs that have to be made or replacement parts to order.)	
❑ Do you see eggs? (You should find only one per cell.)	
❑ Can you find the queen? (Is she the same one you introduced?)	
❑ How's the brood pattern? (It should be compact and plentiful.)	
❑ Evaluate your queen based on her egg-laying ability. (Do you need to replace her with a new queen?)	
❑ How do the larvae look? (Larvae should be a glistening, snowy white.)	

(continued)

Observations	Notations
❑ Check for swarm cells. (Take swarm prevention steps, if needed.)	
❑ Check appearance of brood cappings. (Cappings should be slightly convex and free of perforations.)	
❑ Is the colony healthy? (You should find lots of active bees, healthy-looking brood, a clean hive, and a nice sweet smell.)	
❑ Do the bees have food? (They need honey, pollen, and nectar.)	
❑ How much capped honey is there? (Is it time to add a queen excluder and honey supers?)	
❑ Do the bees have an adequate water supply?	
❑ Clean off propolis and burr comb that makes manipulation difficult.	
❑ Check ventilation. (Adjust based on weather conditions.)	
❑ Is it time to feed? (This usually is done in spring and autumn.)	
❑ Is it time to medicate? (This usually is done in spring, autumn, and when disease is evident.)	

Action Items (What to do between now and the next inspection):

_____

_____

_____

_____

_____

_____

_____

_____

_____

Glossary

• •

abscond: To leave a hive suddenly, usually because of problems with poor ventilation, too much heat, too much moisture, mites, moths, ants, beetles, lack of food, or other intolerable problems.

acarine disease: The name given to the problems bees experience when they are infested with tracheal mites (Acarapis woodi).

Africanized honey bee (AHB): A short-tempered and aggressive bee that resulted from a cross of honey bees from Brazil and Africa. The media has dubbed it "Killer Bee" because of its aggressive behavior.

apiary: This is the specific location where a hive(s) is kept. (Sometimes referred to as a beeyard.)

apiculture: The science, study, and art of raising honey bees. (As a beekeeper, you are an apiculturist!)

Apis mellifera: The scientific name for the European honey bee.

apitherapy: The art and science of using products of the honey bee for therapeutic/medical purposes.

bee bread: Pollen, collected by bees, that is mixed with various liquids and then stored in cells for later use as a high protein food for larvae and bees.

bee space: The critical measurement between parts of a hive that enables bees to move freely about the hive. The space measures ⅜ inch (1 cm).

bee veil: A netting worn over the head to protect the beekeeper from stings.

beehive: The "house" where a colony (family) of honey bees lives. In nature, it may be the hollow of an old tree. For the beekeeper, it usually is a boxlike device containing frames of honeycomb.

beeswax: The substance secreted by glands in the worker bee's abdomen that is used by the bees to build comb. It can be harvested by the beekeeper and used to make candles, cosmetics, and other beeswax products.

bottom board: The piece of the hive that makes the ground floor.

brace/burr comb: Brace comb refers to the bits of random comb that connect two frames, or any hive parts, together. Burr comb is any extension of comb beyond what the bees build within the frames. (Both should be removed by the beekeeper to facilitate manipulation and inspection of frames.)

brood: A term that refers to immature bees, in the various stages of development, before they have emerged from their cells (eggs, larvae, and pupae).

brood chamber: The part of the hive where the queen is laying eggs and brood is being raised. This is typically the lower deep, when two hive bodies are used.

capped brood: The larvae cells that have been capped with a wax cover, enabling the larvae to spin cocoons and turn into pupae.

caste: The two types of female bees (workers and queens) and the male bee (drones).

cell: The hexagon-shaped compartment of a comb. Bees store food and raise brood in these compartments (cells).

cleansing flight: Refers to when bees fly out of the hive to defecate after periods of confinement. (A good day to wear a hat.)

cluster: A mass of bees, such as a swarm. Also refers to when bees huddle together in cool weather.

colony: A collection of bees (worker bees, drones, and a queen) living together as a single social unit.

comb: A back-to-back collection of hexagonal cells that are made of beeswax and used by the bees to store food and raise brood.

crystallization: The process by which honey granulates or becomes a solid (rather than a liquid).

dancing: A series of repeated bee movements that plays a role in communicating information about the location of food sources and new homes for the colony.

deep hive body: The box that holds standard full-depth frames. (A deep box is usually 9⅝ inches deep. It is often simply referred to as a deep.)

drawn comb: A sheet of beeswax foundation used by the bees to build up the walls of the cells.

drifting: Refers to when bees lose their sense of direction and wander into neighboring hives. (Drifting usually occurs when hives are placed too close to each other.)

drone: The male honey bee whose main job is to fertilize the queen bee.

egg: The first stage of a bee's development (metamorphosis).

entrance reducer: A notched strip of wood placed at the hive's entrance to regulate the size of the "front door." Used mostly in colder months and on new colonies, it helps control temperatures and the flow of bees.

extractor: A machine that spins honeycomb and removes liquid honey via centrifugal force. (The resulting honey is called extracted honey or liquid honey.)

feeder: A device that is used to feed sugar syrup to honey bees.

feral bees: Wild honey bees that are not managed by a beekeeper.

food chamber: The part of the hive used by the bees to primarily store pollen and honey. This is typically the upper deep, when two deep hive bodies are used.

foulbrood: Bacterial diseases of bee brood. American Foulbrood is very contagious — it is one of the most serious bee diseases. European Foulbrood is less threatening. Colonies should be treated with an antibiotic (such as Terramycin) to prevent Foulbrood.

foundation: A thin sheet of beeswax that has been embossed with a pattern of hexagon-shaped cells. Bees use this as a guide to neatly build full-depth comb.

frame: Four pieces of wood that come together to form a rectangle designed to hold honeycomb.

hive: A home provided by the beekeeper for a colony of bees.

hive tool: A metal device used by beekeepers to open the hive and pry frames apart for inspection.

honey flow: The time of year when an abundance of nectar is available to the bees.

honeycomb: Comb that has been filled with honey.

inner cover: A flat board with a ventilation hole that goes between the upper hive body and the outer (top) cover.

larva (pl. *larvae*): The second stage in the development of the bee.

laying worker: A worker bee that lays eggs. (Because they are unfertile, their eggs can only develop into drones.)

marked queen: A queen bee that is marked with a dot of paint on her thorax to make it easier to find her, document her age, or otherwise keep track of her.

nectar: The sweet, watery liquid secreted by plants. (Bees collect nectar and make it into honey.)

Nosema disease: An illness of the honey bee's digestive track caused by the protozoan pathogen, Nosema apis. The disease can be controlled with an antibiotic (such as Fumidil-B).

nucleus hive (*nuc*): A small colony of bees housed in a 3 x 5 cardboard or wooden frame hive.

nuptial flight: The flight that takes place when a newly emerged virgin queen leaves the hive to mate with several drones.

nurse bees: Young adult bees who feed the larvae.

outer cover: The "lid" that goes on top of the hive and serves as protection against the elements. (Sometimes called a top cover or telescoping outer cover.)

pheromone: A chemical scent, released by an insect or other animal, that stimulates a behavioral response in others of the same species.

pollen: The powdery substance that is the male reproductive cell of flowers. (Bees collect pollen as a protein food source.)

propolis: A sticky resinous material that bees collect from trees and plants and use to seal up cracks and strengthen comb. It also has antimicrobial qualities. (Also called bee glue.)

pupa (pl. *pupae*): The third and final stage in the immature honey bee's metamorphosis before it emerges from the cell as a mature honey bee.

queen: The mated female bee, with fully developed ovaries, that produces male and female offspring. (There is usually only one queen to a colony.)

queen excluder: A frame holding a precisely spaced metal grid. The device usually is placed immediately below the honey supers to restrict the queen from entering that area and laying eggs in the honeycomb. The spacing of the grid allows foraging bees to pass through freely but it is too narrow for the larger queen to pass through.

queen substance: A term that refers to the pheromone secreted by the queen. It is passed throughout the colony by worker bees.

reversing: The managerial ritual of switching a colony's hive bodies to encourage better brood production. (Usually done in the early spring.)

robbing: The pilfering of honey from a weak colony by other honey bees or insects.

royal jelly: The substance that is secreted from glands in a worker bee's head and is used to feed brood.

scout bees: The worker bees that look for pollen, nectar, or a new nesting site.

shallow super: The box that is used to collect surplus honey. The box is $5^{11}/_{16}$ inches deep. (Sometimes called a honey super.)

smoker: A tool with bellows and a fire chamber that is used by beekeepers to produce thick, cool smoke. The smoke makes colonies easier to work with during inspections.

stinger: The part of the bee's anatomy that everyone knows. The hypodermic-like stinger is located at the end of the adult female bee's abdomen. Remember, bees don't bite! They sting.

supercedure: The natural occurrence of a colony replacing an old or ailing queen with a new queen. (A cell containing a queen larva destined to replace the old queen is called a supercedure cell.)

supering: The act of adding shallow (honey) supers to a colony.

surplus honey: Refers to the honey that is above and beyond what a colony needs for its own use. It is this "extra" honey that the beekeeper harvests for his/her own use.

swarm: A collection of bees and a queen that has left one hive in search of a new home (usually because the original colony had become too crowded). Bees typically leave behind about half of the original colony and the makings for a new queen (queen cells or swarm cells). The act itself is called swarming.

uncapping knife: A device used to slice the wax capping off honey comb that is to be extracted. (These special knives usually are heated electrically or by steam.)

winter cluster: A tightly packed colony of bees, hunkered down for the cold winter months.

worker bee: The female honey bee that constitutes the majority of the colony's population. Worker bees do most of the chores for the colony (except egg laying, which is done by the queen).

Index

• *M* •

• S •

Your #1 Source for Quality Beekeeping Supplies & Equipment

Bee-commerce.com brings you the world's best quality beekeeping supplies and equipment. Our deluxe start-up kits make it easy to get underway! Supplies, tools, medication, honey harvest equipment, books, videos, and bee-related gifts. We even sell package bees and queen bees! Secure online shopping and personalized support designed just for the backyard beekeeper. Most products are shipped the same day we receive your order. And with our 100% satisfaction guarantee, you can order with complete confidence. It's the Internet's first beekeeping superstore!

Mention "Beekeeping For Dummies®" and get a _free_ _gift_ with your order!

SUPER DELUXE STARTUP KIT

This is our most popular and complete startup kit. It's got everything! Simply put, there's no better kit on the market. We guarantee it. Visit us on the web to see our Super Deluxe Kits, plus our other beekeeping kits, supplies, medication, products and bee-related gifts.

Super Deluxe Startup Kit: $370

Includes two deep hive bodies, frames, foundation, inner and outer covers, bottom board, hive stand, smoker, hive tool, veil, leather gloves, hive top feeder, plus a how-to book and video.

HONEY B HEALTHY™

Honey B Healthy is a feeding stimulant that contains 100% pure essential oils (spearmint and lemon grass). These oils stimulate the immune system of your honey bees and help keep them healthy, even in the presence of harmful mites. Just add a teaspoon of Honey B Healthy to your sugar syrup feedings. The 8 ounce bottle is enough for 12 feedings. $12.95

www.bee-commerce.com
Order Online, or call toll free: 1-800-784-1911

Best Quality Supplies and Free Online Advice for the Backyard Beekeeper
Order Online, or Call Toll Free: 1-800-784-1911

FOR DUMMIES®

The easy way to get more done and have more fun

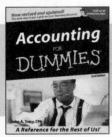

FOR DUMMIES®

A world of resources to help you grow

HOME, GARDEN & HOBBIES

Feng Shui

0-7645-5295-3

Gardening

0-7645-5130-2

Guitar

0-7645-5106-X

Also available:

Auto Repair For Dummies
(0-7645-5089-6)

Chess For Dummies
(0-7645-5003-9)

Home Maintenance For
Dummies
(0-7645-5215-5)

Organizing For Dummies
(0-7645-5300-3)

Piano For Dummies
(0-7645-5105-1)

Poker For Dummies
(0-7645-5232-5)

Quilting For Dummies
(0-7645-5118-3)

Rock Guitar For Dummies
(0-7645-5356-9)

Roses For Dummies
(0-7645-5202-3)

Sewing For Dummies
(0-7645-5137-X)

FOOD & WINE

Cooking

0-7645-5250-3

Cookies

0-7645-5390-9

Wine

0-7645-5114-0

Also available:

Bartending For Dummies
(0-7645-5051-9)

Chinese Cooking For
Dummies
(0-7645-5247-3)

Christmas Cooking For
Dummies
(0-7645-5407-7)

Diabetes Cookbook For
Dummies
(0-7645-5230-9)

Grilling For Dummies
(0-7645-5076-4)

Low-Fat Cooking For
Dummies
(0-7645-5035-7)

Slow Cookers For Dummies
(0-7645-5240-6)

TRAVEL

Italy

0-7645-5453-0

Hawaii

0-7645-5438-7

Las Vegas

0-7645-5448-4

Also available:

America's National Parks For
Dummies
(0-7645-6204-5)

Caribbean For Dummies
(0-7645-5445-X)

Cruise Vacations For
Dummies 2003
(0-7645-5459-X)

Europe For Dummies
(0-7645-5456-5)

Ireland For Dummies
(0-7645-6199-5)

France For Dummies
(0-7645-6292-4)

London For Dummies
(0-7645-5416-6)

Mexico's Beach Resorts For
Dummies
(0-7645-6262-2)

Paris For Dummies
(0-7645-5494-8)

RV Vacations For Dummies
(0-7645-5443-3)

Walt Disney World & Orlando
For Dummies
(0-7645-5444-1)

Available wherever books are sold. Go to www.dummies.com or call 1-877-762-2974 to order direct.

FOR DUMMIES®

Helping you expand your horizons and realize your potential

INTERNET

0-7645-0894-6

0-7645-1659-0

0-7645-1642-6

DIGITAL MEDIA

0-7645-1664-7

0-7645-1675-2

0-7645-0806-7

GRAPHICS

0-7645-0817-2

0-7645-1651-5

0-7645-0895-4

Notes

Inspection Checklists

These tear-out checklists can be used as a quick reference when inspecting your honey bees. You'll find more detailed information in the referenced chapters of the book.

Spring start-up (first inspection of the season)

For more detailed information, see Chapter 8.

❑ As winter crawls to an end, pick the first mild sunny day with little or no wind to inspect your bees (50 degrees F or warmer).

❑ Observe the hive entrance. Are many dead bees around the entrance? A few dead bees are normal, but finding more casualties than that may indicate a problem; see Chapter 10.

❑ Is there *brown spotting* on the hive? These are bee feces, which indicate the presence of nosema disease (see Chapter 10). Even if you don't see the brown spotting, your first spring inspection is time to medicate your bees with Fumidil-B (antibiotic/fumagillin) by adding it to the first gallon of sugar syrup you feed them.

❑ Lightly smoke and open the hive. Do you see the cluster of bees? Can you *hear* the cluster?

❑ Remove a frame or two from the center of the top deep-hive body. Do you see any brood? Look for eggs (eggs mean you have a queen). If you see no eggs or brood, order a new queen from your supplier.

❑ Does the colony have honey? If not, or if they're getting low, immediately begin feeding syrup to the bees.

❑ Feed your colony a pollen substitute to boost brood production.

❑ Add Apistan strips to the brood nest to control varroa mites.

❑ Place a packet of menthol crystals on top of the brood nest to control tracheal mites. Putting this on a small sheet of aluminum foil will prevent the bees from covering the packet with propolis.

❑ Dust the frame's top bars with a mixture of Terramycin (antibiotic) and powdered sugar to prevent foulbrood.

❑ Reverse the deep hive bodies to better distribute the brood pattern. Use this opportunity to clean the bottom board.

❑ Later in the spring, add a queen excluder and honey supers (all medication must be off the hive at this time).

Beekeeping For Dummies®

Cheat Sheet

Routine inspections

For more detailed information, see Chapter 7.

- ❑ Observe the "comings and goings" of bees at the entrance. Do things look "normal," or are bees fighting or stumbling around aimlessly? (See chapters 9, 10, and 11)
- ❑ Smoke the hive (at entrance and under the cover).
- ❑ If you're using a screened bottom board, check the slide-out tray for varroa mites. Clean and replace it.
- ❑ Open the hive. Remove the wall frame and set it aside.
- ❑ Work your way through the remaining frames.
- ❑ Do you see the queen? If not, look for eggs. Finding eggs means that you have a queen.
- ❑ Look at uncapped larvae. Do they look bright white and glistening (normal) or are they tan or dull (indicating a problem; see Chapter 10)?
- ❑ How's the brood pattern? Is it compact (with few empty cells) and does it cover most of the frame? This is excellent.
- ❑ Is the brood pattern spotty (with many empty cells)? Are cappings sunken in or perforated? If yes, you may have a problem (see Chapter 10).
- ❑ Do you see swarm cells? Provide the colony with more room to expand. Check for adequate ventilation.
- ❑ Anticipate the colony's growth. Add additional honey supers *before* it's obvious that the bees need more room.
- ❑ Replace all frames and close up the hive.

Getting ready for winter (last inspection of the season: autumn)

For more detailed information, see Chapter 8.

- ❑ Smoke the hive at the entrance and under the cover as usual.
- ❑ Open the hive for inspection.
- ❑ Confirm that you have a queen. Either find her, or look for eggs.
- ❑ Does the colony have enough honey for its use during the winter? Bees in cold northern states need eight to ten frames of capped honey (less for bees in warm southern states).
- ❑ Feed bees syrup and medicate your colony.
- ❑ Place a sugar-and-grease patty on the top bars of the upper deep.
- ❑ Provide adequate ventilation.
- ❑ Install a metal mouse guard at the hive's entrance.
- ❑ Wrap hive in black tarpaper.
- ❑ Clean, repair and store surplus equipment.
- ❑ Fumigate stored honey supers with paradichlorobenzene (PDB) crystals to prevent wax moth damage (see Chapter 13).

For Dummies: Bestselling Book Series for Beginners